The Cell

The Cell

A Small Wonder

TORSTAR BOOKS
New York • Toronto

TORSTAR BOOKS INC.
300 E. 42nd Street
New York, NY 10017

THE HUMAN BODY
The Cell:
A Small Wonder

Publisher
Bruce Marshall

Art Director
John Bigg

Creation Coordinator
Harold Bull

Editor
John Clark

Managing Editor
Ruth Binney

Text Commissioning
Hal Robinson

Contributors
Arthur Boylston, Tom Cavalier-Smith,
Robert Stewart, Michael Kitch

Text Editors
Wendy Allen, Mike Darton, Martyn
Page, Polly Powell, Sandy Shepherd

Researchers
Maria Pal, Jazz Wilson

Picture Researchers
Kate Duffy, Jessica Johnson

Layout and Visualization
Eric Drewery, Ted McCausland

Artists
Mick Gillah, Greensmith Associates,
Aziz Khan, Mick Saunders

Cover Design
Moonink Communications

Cover Art
Paul Giovanopoulos

Production Director
Barry Baker

Production Coordinator
Janice Storr

Business Coordinator
Candy Lee

Planning Assistant
Avril Essery

International Sales
Barbara Anderson

In conjunction with this series
Torstar Books offers an electronic
digital thermometer which provides accurate body temperature
readings in large liquid crystal
numbers within 60 seconds.

For more information write to:
Torstar Books Inc.
300 E. 42nd Street
New York, NY 10017

Marshall Editions, an editorial group that specializes in the design and publication of scientific subjects for the general reader, prepared this book. Marshall has written and illustrated standard works on technology, animal behavior, computer usage and the tropical rain forests which are recommended for schools and libraries as well as for popular reference.

Series Consultants

Donald M. Engelman is Professor of Molecular Biophysics and Biochemistry and Professor of Biology at Yale. He has pioneered new methods for understanding cell membranes and ribosomes, and has also worked on the problem of atherosclerosis. He has published widely in professional and lay journals and lectured at many univer- sities and international conferences. He is also involved with National Advisory Groups concerned with Molecular Biology, Cancer, and the operation of National Laboratory Facilities.

Stanley Joel Reiser is Professor of Humanities and Technology in Health Care at the University of Texas Health Science Center in Houston. He is the author of *Medicine and the Reign of Technology*; coeditor of *Ethics in Medicine: Historical Perspectives and Contemporary Concerns*; and coeditor of the anthology *The Machine at the Bedside*.

Harold C. Slavkin, Professor of Biochemistry at the University of Southern California, directs the Graduate Program in Craniofacial

Biology and also serves as Chief of the Laboratory for Developmental Biology in the University's Gerontology Center. His research on the genetic basis of congenital defects of the head and neck has been widely published.

Lewis Thomas is Chancellor of the Memorial Sloan-Kettering Cancer Center in New York City and University Professor at the State University of New York, Stony Brook. A member of the National Academy of Sciences, Dr. Thomas has served on advisory councils of the National Institutes of Health.

Consultants for The Cell

Robert E. Pollack is Professor of Biological Sciences and Dean of Columbia College at Columbia University. He is also Visiting Professor of Pharmacology at the Albert Einstein College of Medicine. He is a member of the Scientific Advisory Board of the National Alzheimer's Disease Foundation and editor of student texts at the New York Academy of Sciences. Professor Pollack has published many articles on cytology, virology and molecular biology, and is currently engaged in cancer research.

Jean Paul Revel is the A.B. Ruddock Professor of Biology at the California Institute of Technology (Caltech), a former Professor of Anatomy at Harvard Medical School and past president of the American Society for Cell Biology. He has been a member of a number of national committees, and sits on the Editorial Boards of several

scientific journals. His main areas of research are communication between cells and cellular structure, which he studies using cytological and molecular approaches.

Medical Advisor
Arthur Boylston

**Library of Congress
Cataloging in Publication Data**
Main entry under title:

The Cell: a small wonder.

Includes indexes.
1. Cytology.   [DNLM: 1. Cytology—
popular works.   QH 582.4 C393]
QH581.2.C417 1985   611'.018 85-4805
ISBN 0-920269-62-1

ISBN 0-920269-22-2 (The Human Body series)
ISBN 0-920269-62-1 (The Cell)
ISBN 0-920269-63-X (leatherbound)
ISBN 0-920269-64-8 (school ed.)

20 19 18 17 16 15 14 13 12 11
10   9   8   7   6   5   4   3   2   1

Printed in Belgium

Contents

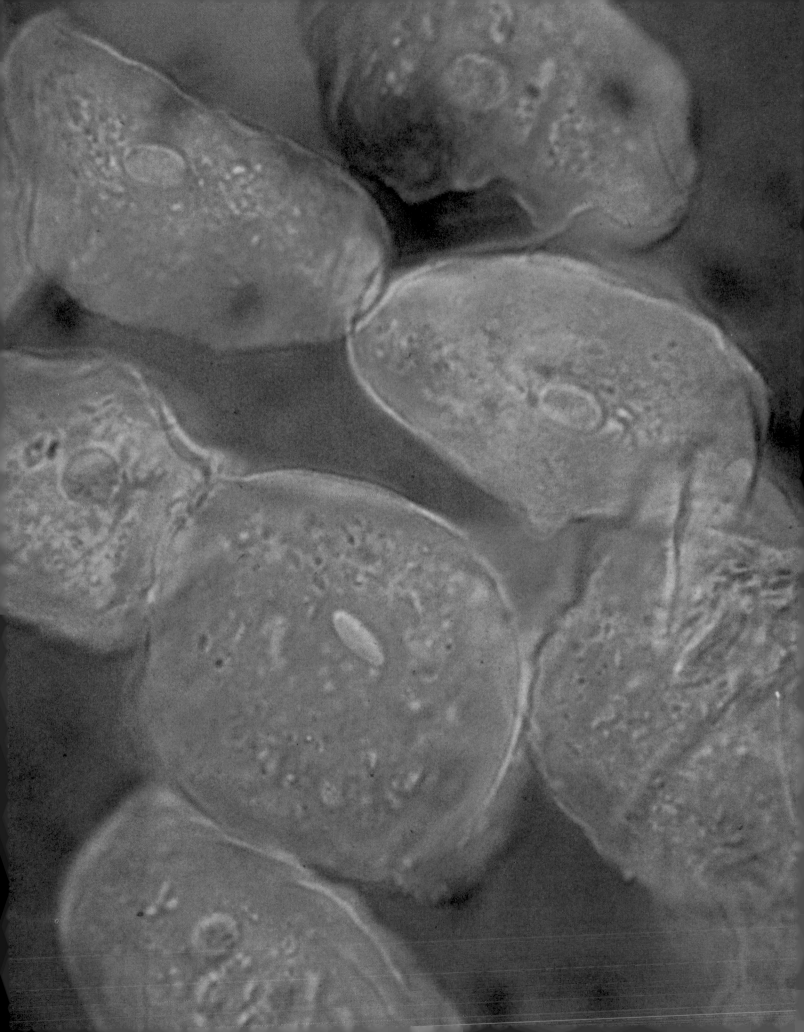

Introduction:

Life's Little Secret

In many ways it is entirely appropriate that the English word "cell" derives from a linguistic root that essentially means "to hide," "to conceal." For more than two millennia the cell was nature's little secret, its form and contents—even its existence—quite beyond the scope of science. "Nature loves to hide," said Heraclitus, and although other ancient Greeks — especially Democritus — contributed to the idea that all matter was made up of particles, it was not until the early nineteenth century (A.D.) that the development of microscopes permitted scientists to actually observe cells and to formulate theories about their structure and how they operate.

Now we know more.

Everyone begins life as a single cell, a fertilized ovum, in size just visible to the naked eye. By the time of our adult maturity, our cell-count totals several hundred trillion, and most cells are much smaller than an ovum. But among that host of cells, some three billion of them—mostly in the surface of the skin—die every 60 seconds. Virtually all are instantly replaced through cell division among those that remain. The total number of cells in the body thus remains relatively constant until old age causes a natural decrease to set in.

Cytology (the study of cells) has burgeoned as a science, followed closely by histology (the study of tissues). The cell's structure and contents have been identified and their significance, with a few reservations, generally appreciated. The relationship between the nature, presence or absence of particular cells and various diseases, disorders and deformities to which human beings are prey has been accorded considerable attention. Morever, the advance of knowledge has enabled scientists even to create new and different kinds of cells, with properties that are beneficial or commercial: the biochemical industries are expanding fast. Much is known; what remains to be discovered could be equally significant.

A group of cells cluster together to make up the soft, subtle tissue that gives form to the contours of a human face. Individually, each cell is a miraculous minute biochemical factory, producing organic substances to maintain itself and serve hundreds of functions throughout the body.

Chapter 1

The Search for Life

It is sometimes thought that the biological sciences began only a couple of centuries ago as scientific methods became more sophisticated, allowing scientists to study the structures of animals and plants, and investigate how they lived. Yet the true origins of biology lie much earlier, obscured by the mists of time. The earliest records are from the sixth century B.C., attributed to the great pioneering student of nature Anaximander, who lived in the Ionian Islands, part of the ancient Greek community on the Mediterranean.

Anaximander, a pupil and friend of the renowned philosopher Thales of Miletus, was the first man in recorded history to consider the whole of nature with reasoning curiosity, and to set down his findings. Almost everything he wrote has been lost, but the pieces that do survive are remarkable.

Imagination is one source of scientific inventiveness, and Anaximander's mind was of amazing power and fertility. He even developed a concept of evolution. His main idea was that evolution occurred by the separation of complementary opposites — so that, for example, living ("dry") creatures arose from a moist element, the sea. Other complementary constituents of the universe included heat and cold. Anaximander thought that the first human came into being as a fish from the inside of other fishes, living in water until eventually he was cast ashore and forced to adapt to life on dry land.

Anaximander's conclusions may seem fanciful now, but what makes them interesting still is the manner in which they were reached: no blind acceptance that the world and everything in it was permanent, no recourse to the supernatural, no shrinking from reproof by contemporary religious authorities stood in the way of his attempt to accumulate a genuinely rational corpus of knowledge about the observable world around him. The only problem was, of course, that the "data" he used to build his view of the world was fanciful, not

In the thirteenth-century depiction of the creation on the vault of the atrium of St Mark's Cathedral, Venice, the central circle represents an omnipotent creator. The three outer rings show the creation of the world, the origin of man, and man's subsequent history. Such images exemplify man's powerful desire to understand, and impose order on, the natural world in the search for the key to life — in this case by invoking an all-powerful entity as the ultimate controller.

9

rooted in fact. It is likely that even the earliest human beings puzzled over questions about the mystery of life. But the study of biology probably started with men of vision such as Anaximander — who began to seek answers by trying to remove the mystery and to scientifically investigate life itself.

Every important Greek thinker was a polymath, desiring to learn everything about everything. One of the consequences of that universality was that elements which today would be called "scientific" could be woven naturally into more speculative thought of the kind that is now categorized as "philosophical." What is called biology was regarded as part of "natural philosophy." The factual ideas of the ancients are thus often difficult to disentangle from their other thoughts.

Yet in an age when Homer's epic heroes were mere playthings of the gods, their fortunes and misfortunes often the result of divine intervention and therefore miraculous, the scientific search for operative causes went on. Those men who believed that events, not the gods, were the cause and who had the tenacity and the courage of their convictions, slowly learned how to investigate nature. Much of the impetus came from medicine, from the desire to understand the workings of the human body in order to treat disease and injury. Greek medicine was, however, primitive. Hippocrates, for instance, subscribed to the prevailing theory that four "humors" constituted the elements of the body: blood was the warm/moist element; yellow bile (choler) the warm/dry; black bile (melancholy) the cold/dry; and phlegm the cold/moist one. The theory had no "factual" foundation, but Hippocrates' concept that illness arose from a disorder in the relationships between the humors was an advance. It recognized the organic unity of the body and attempted to locate the origin of disease in natural causes. Greek physicians no longer thought that ill-health could be warded off by consulting the stars, or disease cured by propitiating the gods.

Plato: A Hindrance to Progress

For physical events there have to be physical explanations, and for physical objects there have also to be physical descriptions. These facts appear obvious now, but were not so to the Greeks. Plato (*c.* 427–347 B.C.) based his philosophy on the fundamental argument that reality lay in the realm of abstract thought. What appears to be real — taken to be real by the evidence of the senses — was for Plato merely the imperfect, temporal, earthly manifestation of the eternal ideal. Plato has his place in the early beginnings of biology, for his very notion of the "idea" of a horse, or the "idea" of a bone, had important consequences, leading to an awareness of species as opposed to individual animals or objects. His followers laid the foundations for the classification of living things into related groups (like the modern genera and species).

In other directions, however, Plato exercised an obstructive influence on the natural sciences. Since thought rather than practical experiment was his ideal source of knowledge, what could be the point of examining a fish? The real thing was simply an imperfect representation of the ideal of "fishness." Occasionally Plato's conclusions were amusing. In the dialogue *Timaeus*, for example, in which he propounds his theory of the origin of the universe, he explains how human beings come to have the shape that they do. His explanation is based on the notion that the sphere is the most perfect celestial form. The human head, which he thought gave lodging to the soul, is therefore an approximation to the sphere, and is all that is necessary — except that to prevent it from rolling on the ground the gods added a trunk and limbs.

The Atomists

Another important group of Greek thinkers — who may be loosely banded together under the term "atomists" — arrived at a pragmatic outlook that has provided the basis for scientific investigation to the present day. Their work, especially that of Democritus (*c.* 460–370 B.C.), reasserted Anaximander's contention that the origins of life and the characteristics of matter must have a natural explanation. Democritus, who came from the Greek colony of Abdera, had as a teacher the philosopher Leucippus and inherited from him the theory that the universe consists of particles moving in space (a theory at which his contemporary, Zeno of Elea, also hinted in his instructional

paradoxes). Only a few fragments of Democritus' vast literary output survive, but they are sufficient to establish him as the founder of the atomic theory, one of the simplest and most fruitful theories in the history of science.

The theory as Democritus stated it may be outlined briefly. The universe consists of two fundamental realities: space (or the void) and atoms. Everything else — heat and cold, for example — has no true reality. Because nothing can be generated out of nothing, nor anything which exists cease to exist, all change is nothing more than a rearrangement of atoms, a change in the compound forms that aggregations of atoms assume. Furthermore, all events happen only through causes and physical necessity. Nothing happens without cause and nothing happens by intention (that is, a particular event cannot be made to occur by an act of will).

Following these ideas Democritus succeeded in providing some truly remarkable insights. His view that motion continues unless something stops it

anticipated the ideas of Newton. His notion that atoms had different sizes and shapes bears a surprising resemblance to modern molecular theory, and led him to suggest that atoms of a liquid were round and smooth and therefore unable to attach themselves to each other, whereas those of a solid were jagged and angular so that they gripped onto each other.

Democritus sought to explain human life by distinguishing between bodily or "corporeal atoms" and "soul atoms." The corporeal atoms were fixed in the parts of the body to which they belonged. Soul atoms, similar to those that made up the basic element of fire, were lighter and moved about the body, giving it life. Because these "fire atoms" were extremely light and mobile, they were constantly being given off by the body, which continued to live only by inhaling fresh supplies of them. When respiration ceased, the body died. Sleep was explained as the result of the loss of fire atoms, although not in sufficient quantity to cause death.

Democritus was the first of the Greeks to locate thought in the brain and to insist upon the complexity of organisms in the subhuman animal

kingdom. His eminence, however, derives less from his particular conclusions than from the scientific approach of his mind. Democritus substituted the idea of necessity for those of chance and predetermination. Once atoms were set in motion, everything followed by immutable and ascertainable laws. It was the business of natural philosophy to try to discover those laws. Modern science is based on this simple concept.

That doctrine of necessity, arising from a pragmatic interpretation of the physical universe, was the chief legacy that René Descartes in the seventeenth-century (A.D.) and the nineteenth-century materialists inherited from Greek thought. In the ancient world it was given its most memorable and most detailed expression in Lucretius' long poem *De rerum natura* ("On the Nature of Things"). Lucretius (*c.* 95–55 B.C.), a Roman poet and philosopher, seems to have come by his pragmatic outlook through studying the works of Epicurus (who wrote three centuries earlier). Lucretius was an ardent popularizer. Having adopted the central idea of atomism with all the passion and zeal of a convert, he wrote his poem — he said — in order to explain to his fellow countrymen in "the seducing accents of the Muses" the "obscure discoveries of the Greeks." Apart from the works of Aristotle, no survival from antiquity did more than *De rerum natura* — which first appeared in printed form in Europe in 1473 — to inspire the revival of science which accompanied the Italian Renaissance.

Aristotle

By universal agreement it is the writings of Aristotle (384–322 B.C.) that constitute the crowning glory of Greek science. Aristotle laid the foundations of comparative anatomy. He dissected hundreds of specimens; he examined the three-day-old egg of a chicken and saw the tiny heart beating and the blood vessels spreading out over the yolk. He divided animals into those that had blood (which, to him, had to be red) and those that did not — a fundamentally trivial distinction, but an imaginative and useful step toward a general system such as the modern classification into vertebrates and invertebrates. His differentiation between cartilaginous and bony fishes still stands

The ancient Greek thinker Democritus made contributions to virtually all areas of knowledge, although probably best known for his atomic theory. This he applied both to physical phenomena and biology.

According to Democritus' atomic theory, atoms of a solid (top left) were angular so that they interlocked firmly, whereas those of a liquid (top right) were round and smooth and were therefore unable to attach

themselves to each other. We now know that all atoms are spherical and that the difference between a gas (bottom right), a liquid (bottom middle) and a solid (bottom left) is in the motion and bonding of atoms.

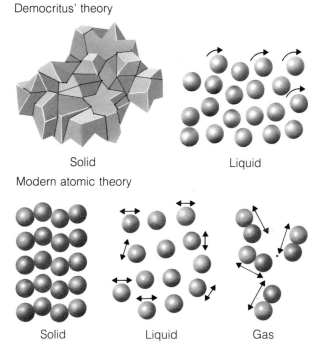

Democritus' theory

Solid   Liquid

Modern atomic theory

Solid   Liquid   Gas

in zoological classification. Aristotle was scornful of his contemporaries who deemed it beneath their dignity to carry their observations beyond the world of humans: "If any person thinks the examination of the rest of the animal kingdom an unworthy task, he must hold in like disesteem the study of man. For no one can look at the primordia of the human frame—blood, flesh, bones, vessels and the like—without much repugnance."

The range of Aristotle's investigations was unprecedented, as were the lengths he went to in his attempts to bring order to the chaos of his observations. He recognized the distinction between homology (resemblance in structure and development) and mere analogy (resemblance in use or function), and he understood that the biologist cannot simply study the remains of dead specimens but must examine the ongoing processes of living ones. The study of living organisms inevitably brings the biologist to a consideration of the larger form, and how it is composed of its parts. Architecture, Aristotle pointed out, could not be understood simply by knowing what bricks, mortar and timber were. Neither could living creatures. "When any one part

or structure is under discussion, it must not be supposed that its material composition is the object of the discussion, but rather the relationship of that part to the total form." It was not enough merely to isolate and describe the various substances of which the body was composed, such as flesh, blood and bone; it was more important to understand the relationship of parts, of heart to liver, or of face to hands. "For we should not be content to say that the couch was made of bronze or wood, but should try to describe its design or mode of composition in preference to material . . . For a couch is such-and-such a form embodied in this or that matter."

Aristotle amazes many readers by the modernity of his insights. A particularly striking passage reads: "Of the parts of animals some are simple: namely, all such as divide into parts uniform with themselves, as flesh into flesh; others are composite, such as divide into parts not uniform with themselves, as, for instance, the hand does not divide into hands nor the face into faces." The distinction was sound and apt, comprehensible to Aristotle's contemporaries. But its real validity was recognized only centuries later when microscopic investigation became possible.

Raphael's fresco The School of Athens *in the Vatican Palace shows Plato and Aristotle surrounded by philosophers and thinkers from many ages, including Socrates, Ptolemy, Heraclitus, Euclid, Zoroaster and Epicurus. Painted in 1510–1511, in response to the artistic and scientific ferment in Renaissance Italy, Raphael's work illustrates the renewed interest in ancient Greek — particularly Platonic — thought.*

Another passage runs: "It is evident that there must be something or other really existing, corresponding to what we call by the name of Nature. For a given germ does not give rise to any random living being, nor spring from any chance one, but each germ springs from a definite parent and gives rise to a predictable progeny. And thus it is the germ that is the ruling influence and fabricator of the offspring." Making an allowance for the difference of terminology, it is possible to see that here, in the fourth century B.C., there is more than a suggestion of Mendelian genetics and modern cell theory.

After Aristotle

How was it, then, that centuries passed before Aristotle's wisdom became common knowledge? What kept it hidden in darkness for nearly two thousand years?

Christianity is often held to be responsible, and certainly throughout its history the Christian Church has, with fluctuating intensity, placed itself in opposition to purely rational contemplation of the physical universe. At the same time it should be remembered that the scientific vein of the ancient world had practically dried up following the death of the Greek physician Galen in the second century A.D., long before the Christian faith had established itself in Europe. The effects of Plato's love of idealization were partly responsible for the decline of science. The spread of the "barbarians" from the north, who carried no tradition of scientific or intellectual achievement with them, were also partly to blame.

Scientific culture requires discussion and exchanges; until the rise of the cities from about the twelfth century, everyday rural life in Europe was generally unsuited to the growth and cross-fertilization of new ideas. In the meantime, the Greek heritage was preserved and augmented by the Jews — such as the physician and philosopher Moses Maimonides — and the Muslim Arabs who

Aristotle

The First Biologist

One of the greatest philosophers of all time, Aristotle was praised by contemporaries and successors alike. His outstanding contribution to philosophy is undisputed but his investigations into other fields, particularly those of the natural sciences, are of equal importance.

He was born in 384 B.C. in Stagira, a colony of northern Greece, close to the kingdom of Macedonia where his father was physician to Amyntas II. Orphaned at an early age, Aristotle was brought up by a friend of the family. He had spindlelegs and a lisp, physical defects for which he compensated by dressing well and sporting jewelry. He was probably something of a dandy.

Drawn to the intellectual climate of Athens, Aristotle studied at Plato's Academy where the great master, who had just finished writing *The Republic*, referred to him as "the intelligence of the school." On Plato's death Aristotle left Athens, and for a time lived on the Asiatic mainland at Assos, where he drew around him other great thinkers, and wrote *On Philosophy* in which he attempted to define civilization.

Aristotle's move to the island of Lesbos brought with it a directional change in interest and he began to devote himself with passion to biological research. He attempted to classify and make a hierarchical system of animal species ("the ladder of nature") and was particularly interested in the creatures of the sea. He classified dolphins with some land animals because he observed that they gave birth to live young and nourished the fetus through a placenta. This theory was scorned by his successors and only recognized as correct in the nineteenth century. Needless to say, he was not always right in his assumptions; he believed the heart to be the center of all life and regarded the brain as a simple cooling chamber.

In placing animals in a primitive hierarchy Aristotle's findings naturally led him towards the idea of a sort of evolution whereby there existed a chain of change.

From biological experiments, Aristotle went on to ponder on the relationship between the soul and the body. He deduced, in contrast to Plato, that the affects of a soul are inseparable from the material of the body. The soul is the Form of the body and body is the Matter of the soul. Accepting that immortality resides only in the species and not in the individual, he was approaching the notion that life is within biological boundaries.

For a time, Aristotle tutored Phillip II of Macedonia's son who was later to become Alexander the Great. At the age of 50 he returned to Athens and established the Lyceum, an alternative school to the Academy. On the death of Alexander the Great in 323 B.C. however, Aristotle left Athens because he feared an anti-Macedonian reaction amongst Athenians, saying that he would not allow the city to "sin twice against philosophy" (referring to the fate of Socrates). He died one year later in his mother's hometown of Chalcis.

replaced the Romans as the dominant power in Asia, northern Africa and southern Europe. In the eleventh century the Arab physician Avicenna revived the works of Aristotle, which led eventually to their being translated from Arabic into Latin for the libraries of Italy. Yet the reappearance of Aristotle's texts in Europe could initially be no more than potential inspiration, because it was one of the shortcomings of the scholarship of the ancient Greek scientists that they published their results but scarcely ever recorded the methods they used. Most biological investigation, when it was taken up again in the fifteenth century (especially by Italian sculptors and painters interested in anatomy), had to start once more from the beginning, but nevertheless had the benefit of the experimental approach to science initiated by Galileo Galilei (1564–1642).

Descartes and the Reawakening of Science

There were many strands in the reawakening of the investigative spirit that produced the scientific revolution of the seventeenth century. One man, nevertheless, may be said to have summed up the scientific tendencies of early modern Europe and to have given them an ordered expression which proved capable of serving as the foundation for the subsequent progress of Western science. He was René Descartes (1596–1650), the French philosopher, mathematician and biologist. Descartes' philosophy rested on the assumption that "I think, therefore I am." He is sometimes represented as being thoroughly uninterested in experimental work, although that picture is far from true. He dissected embryos of birds, cattle, fish, cats and dogs. He regularly instructed his local abattoir to send him livers, hearts and eyes. He observed the working of the human nervous system, of reproduction, and following the work of the contemporary English doctor William Harvey, of the circulation of the blood. And he did all this in the spirit of the Greek pragmatists. The human body, in his words, was a *machine de terre*, an earthly machine of which the characteristics and functions were to be understood and described in terms of his intuitive notion of truth.

Descartes did not deny the existence of the soul; he insisted, rather, on the duality in man of soul and body. But the soul — whatever it was — was not the proper study of the scientist. The biology of man could be described entirely without reference to it. "I have described," Descartes once said, "the whole visible world as if it were only a machine in which there was nothing to consider but its shapes and movements." This was similar to the pragmatism of the ancients, but contrary to the more scientific experimentation of men such as Galton and Galileo.

So Descartes revived and extended the intellectual framework of the ancients. Yet it would be wrong to trace a strict causal continuity between Greece and the modern world in the genealogy of science. Modern etiology arose from the anatomical studies made by men such as Leonardo da Vinci (1452–1519) and (in the next generation) Andreas Vesalius, and — above all — from the invention of the microscope.

The Cell Brought into Focus

The power of spherical pieces of glass to magnify objects had been known to the ancients. Ptolemy wrote a treatise on optics in the second century A.D. describing the phenomenon. Even so, it was not until the late sixteenth century that the history of microscopy really began, when the placing of a tube between two lenses created the Galilean microscope.

Galileo was not the actual inventor of the microscope. Whether the credit is due to any one person — such as Zacharias Janssen of Holland, who made one around the year 1590 or, as is more probable, to a number of spectacle-makers throughout western Europe is not clear. At any rate, Galileo constructed his first telescope in 1608 and, by rearranging the lenses of his telescope, found that he could magnify close objects, so that this new instrument was, in effect, a microscope. Thirty years later Descartes published a sketch of a rather unwieldy apparatus, which he called the "ideal microscope." Its chief feature was the use of two lenses, one at each end of a tube — in other words, it was a compound microscope.

Two of the most interesting early microscope-makers were Athanasius Kircher and Antonie van Leeuwenhoek. Kircher, a German Jesuit priest, provided the first authentic observations of

Philosopher and scientist René Descartes (the rightmost figure) had a profound influence on science. He believed in mechanistical explanations for many biological and physical phenomena.

microscopic organisms in 1658 when he saw "worms" in decaying substances, and "corpuscles" in the blood of plague victims. Kircher's microscope consisted of a short tube with a lens at one end and a flat glass disk at the other; the object to be examined was placed on the glass disk and viewed through the lens. This simple microscope was initially used to look at insects, and became known as a flea-glass or fly-glass. The tube was only about the size of a man's thumb.

The instruments of the Dutch lens grinder, van Leeuwenhoek, were more sophisticated because of the excellent lenses he made. Still using single, hand-ground lenses he was able to achieve magnifications of up to about 270 diameters. One of his original instruments, presently housed at the University of Utrecht in the Netherlands, makes use of a lens held in place between two perforated copper plates, fitted with two screws for moving the observed object closer to or farther away from the lens. The specimen could be illuminated either by direct transmitted light or by means of a concave reflector. When van Leeuwenhoek was observing

small fish or tadpoles, he would place the specimens in water in a slender glass tube held in a metal frame, to which was attached a plate carrying the magnifying glass. His work produced several significant results, among which were the observation of free, independent cells (although he did not name them as such) which were in fact red blood corpuscles, of spermatozoa and yeast cells, of muscle cells in meat, and the discovery of single-celled animals now known as protozoa.

The Storerooms of Life

At about the same time as van Leeuwenhoek was discovering his free, independent cells, others were making similar discoveries. The honor of first using the word "cell" in a biological sense (the word derives from the Latin *cella*, "storeroom" or "small container") belongs to the English naturalist Robert Hooke (1635–1703).

In his *Micrographia* of 1665, he described the structure of cork as consisting of numerous cavities similar to those found in a honeycomb. The Italian microscopist Marcello Malpighi observed red blood cells — although they may have been seen a few years earlier by the Dutchman Jan Swammerdam. The imprecise dating of both men's observations makes it difficult to assign priorities.

Neither Hooke nor Malpighi, whose descriptions of cells referred to their outer casings, understood that it was the inner contents of the cells that really mattered (a Danish contemporary called them "bladders" which implies that he saw them as containers). Nor did the various seventeenth-century observations of corpuscles and tissue inspire more than general theories until the early nineteenth century. Progress was impeded by limitations in the optical quality of lenses. Those of the compound microscope suffered from chromatic aberration, a tendency to break up white light into its constituent colors. This was not overcome until the invention of the achromatic lens in the 1820s.

The delay was not, however, entirely due to technical inadequacies; a major theoretical barrier had also to be overcome. What held cell theory back was the universally accepted belief that organs, in both plants and animals, were the source of all biological functions. This belief predominated until 1802 when the French anatomist Xavier Bichat, in one of those flashes of insight that illuminate the history of science, realized that it was not the organs but the structure of the tissues of which they were made that governed the physiological processes of living matter. Unfortunately, Bichat died the same year, but he has since been credited with contributing greatly to the foundation of the science of histology, the study of tissues.

Appropriately, it was in that very same year, 1802, when cell theory was on the brink of discovery, that the word "biology" made its appearance in the world of science. Rudolf Treviranus, a German theorist, used it in the title of a book, and the illustrious biologist Jean-Baptiste de Lamarck used it (presumably quite independently) in his *Recherches sur l'organisation des corps vivants* ("Studies in the Organization of Living Bodies") to signify the unity of all branches of study of living things. Lamarck came close to identifying cell division as the fundamental activity of all living matter. "Every step which Nature takes when making her direct creations," he wrote in 1809, "consists in organizing into cellular tissue the minute masses of viscous or mucous substances that she finds at her disposal under favorable circumstances." To say this, however, was a sign perhaps that Lamarck, for all his inventiveness, was in the matter of cell theory still on the other side of the line dividing pre-modern from modern biology.

In 1824 René Dutrochet was one of several contemporary European biologists who independently arrived at the conclusion that "the cell is the fundamental element in the structure of living bodies, forming both animals and plants through juxtaposition." Then in 1831 the botanist Robert Brown — who is probably best known for his discovery of "Brownian movement" (the random and continual movement of particles in a liquid or gas) — made the first exciting observations of what went on inside a cell. Charles Darwin called on him shortly before his own voyage on the *Beagle* and was invited to peer through a microscope and describe what he saw. "This I did, and believe now," Darwin wrote later, "that it was the marvellous currents of protoplasm in some vegetable cell. I then asked him what I had seen; but he answered me, 'That is my little secret!'."

Antonie van Leeuwenhoek, of Delft, Holland, was a cloth merchant by trade. In his spare time he ground lenses for his microscopes, which were unsurpassed until the nineteenth century.

One of van Leeuwenhoek's original microscopes is shown above; with crude-looking instruments such as this he observed red blood corpuscles, spermatozoa and the minute single-celled organisms called protozoa.

Robert Hooke's microscopical apparatus is shown here arranged to view specimens by reflected light. The apparatus could also be used to examine objects by transmitted light, by tilting the microscope so that it was horizontal. Using this equipment, Hooke made several important observations, which he published in his famous work **Micrographia** *(1665). In this book he illustrated the honeycomblike structure of cork and coined the term* cell *to describe the individual units of the "honeycomb." He also described in detail the structure of feathers, the stinger of a bee, the radula ("tongue") of mollusks, and the foot of a fly.*

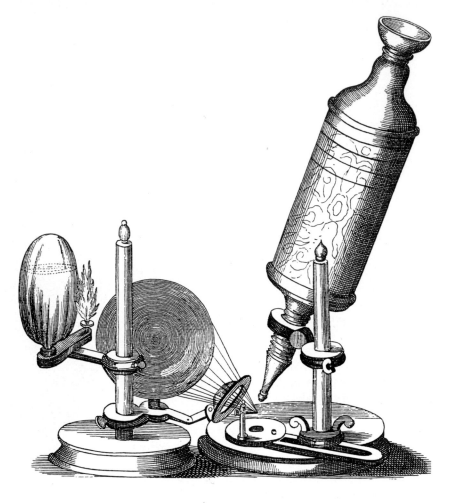

Brown's secret, announced to the world in 1831, was to have detected substructures within a cell. The 1830s saw an explosion of cell observation, especially at the Universities of Breslau (now Wrocław in Poland) and Berlin in Germany. From the former came Johannes Purkinje's systematic studies of animal tissue, particularly of bone cells and nerve cells (he introduced the term "protoplasm"), and G. G. Valentin's discussion of cells in his handbook on embryology. At Berlin the rival school was led by Johannes Müller, who made his name analyzing cartilage and glandular tissue. He also devised a theory of specific nerve energy, and his pupils and disciples made the first observations of cell division.

Müller's researches in physiology and pathology in Berlin forged a link between experimental biology and medicine. For example, he made use of a microscope to study the cellular structure of various kinds of tumors. Later he turned to collecting and classifying specimens of animals, particularly lower forms of invertebrate sea creatures such as protozoans, sea anemones and sea urchins. Through his various researches —

and particularly through the continuation of his principles by students who had worked with him — Müller turned the former mere speculation about nature into a systematic method of scientific observation and experimentation.

One of Müller's pupils was the anatomist Theodor Schwann (1810–1882) and, although it is difficult to credit a single individual with the first formulation of "cell theory," it was Schwann who coined the term in 1839. It is also traditional to date the birth of the theory to a famous meeting between him and Matthias Schleiden (1804–1881) a year earlier. Schleiden was a lawyer who quit the bar to devote himself to the study of physiology and the structure of plants. In October 1838 they met at Schwann's laboratory, and from their discussions came certain conclusions about the similarity between plant and animal tissue. Before this there had been independent observations only partially connected. Now there was an attempt to bring an underlying theoretical unity to biology.

In 1838 Schleiden stated that the lower animals consisted of one cell, the higher animals of many cells. In the following year Schwann declared that

Despite the relative crudity of their instruments, the early microscopists made some remarkable observations —as this eighteenth-century illustration of various "animalcules" shows. For example, illustrations 19–21 depict several aspects of the ciliated protozoan Vorticella and 29–40 are a series of observations of spermatozoa (31–36 are human sperm), redrawn from van Leeuwenhoek's own studies.

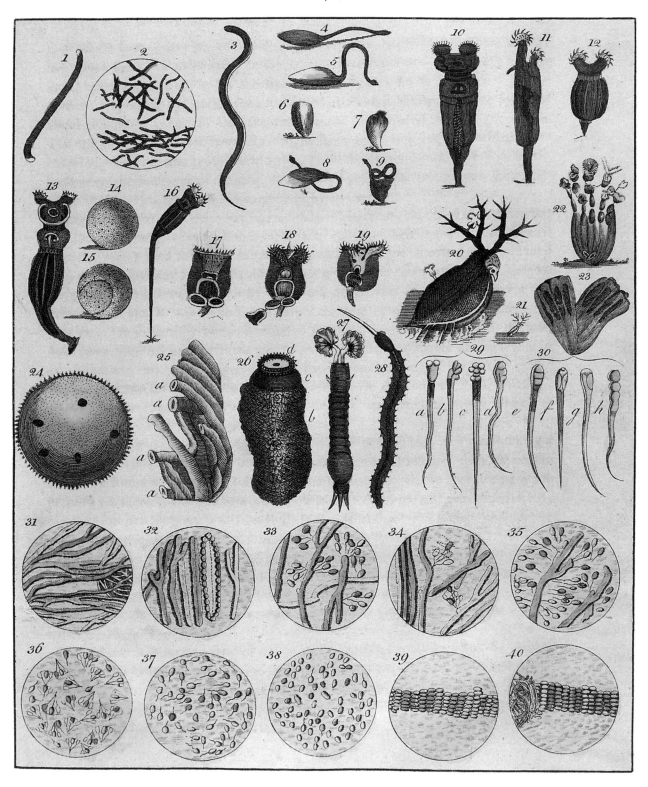

Jean-Baptiste de Lamarck

Pioneer of Modern Biology

Described at the end of the nineteenth century as the "one great mind, who far more than all others combined paved the way for the new science of biology to be founded by Darwin," Jean-Baptiste de Lamarck died blind and penniless, ignored by the world around him. Throughout his life personal tragedy was never far away: he lost four wives and three children. A reassessment of his work since his death, however, has assured that he will always be remembered as one of France's great naturalists.

He was born on 1 August 1744 in Bazentin, a town in northern France, the eleventh child of impoverished aristocrats. His parents envisaged a career for him in the Church, against his wishes, but on the death of his father he was free to join the army. He received an officer's commission for bravery during the Seven Years' War but before long ill health forced him to resume civilian life.

The work for which Lamarck is probably best remembered is his *Zoological Philosophy* published in 1809. In this book he puts forward a theory of evolutionary development whereby species are constantly changing and progressively developing.

The conclusions that Lamarck reached were the result of lengthy studies into plants and, particularly, animal life. During the war he had been stationed on the Mediterranean coast where his interest in the natural world blossomed. A book he published on French flora gained him the attention of other naturalists and established his reputation, which was given the seal of royal approval in 1781 when he was appointed botanist to the King. This position enabled him to travel widely in Europe, where he visited museums and collected minerals as well as animal and plant life.

As professor of invertebrate zoology at the Museum of Natural History in Paris, at the age of 50 Lamarck reached the pinnacle of his career, and it was during this period that his greatest achievements were made. He set about ordering the invertebrates, including the eight-legged spiders, ticks and scorpions in a class of their own (arachnids) and establishing a category for the crustaceans. His findings were summarized in the seven-volume work entitled *Natural History of Invertebrates*, which is commonly acknowledged as the basis of modern invertebrate zoology.

The discovery of giraffes provided Lamarck with what he thought to be a prime example of the effects of evolutionary change. He hypothesized that their ancestors could have been primitive antelopes that had stretched their necks, legs and tongues in the search for young shoots at the tops of trees. Successive generations eventually evolved into the giraffe, through the "inheritance of acquired characteristics." Although his findings may have been entirely incorrect, they stimulated interest in evolutionary theory. But the popularity of Cuvier's nonevolutionary theories caused Lamarck's fall from grace; he died in Paris in 1829.

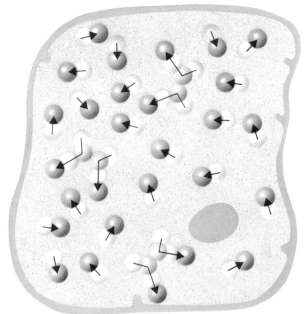

the basic constituent of all living matter, animal or vegetable, was the same. "The elemental parts of the tissue are the cells, similar in general but diverse in form and function. It can be taken for granted that the cell is the universal mainspring of development and is present in every type of organism. The essential thing in life is the formation of cells." Schleiden, in his turn, recognized that protoplasm (whose behavior was first described by Hugo von Mohl in 1845) and the nucleus were the important parts of the cell, while Schwann recognized the importance of a cell wall or membrane in protecting and determining its contents.

In a famous Latin phrase, *omnis cellula e cellula*, the German pathologist and statesman Rudolf Virchow (1821–1902) propounded a new age in cell theory in 1858. His pronouncement that "every cell comes from a cell" was in sharp contrast to the belief of many biologists — including Schwann — that cells originated by some process of coagulation from liquids. The understanding of cell division spread quickly, however, and led to entirely new lines of investigation into how cells multiply.

Within twenty years or so the key processes of mitosis (cell replication, first described by Walther Flemming) and meiosis (cell division to produce the germ cells, noted by Oskar Hertwig, first observer of the process of fertilization) were being studied in a variety of specimens. And almost daily there came fresh evidence that all life did spring from cells and did indeed consist of cells. Virchow, by his pathology research, established that injury to the body was repaired by cell growth and that death was caused by the death of individual cells, in turn causing the death of the whole body.

Improved Resolution and Stains

In the hundred or so years since the development of the modern microscope the biological outlook has been transformed, and has produced a wealth of detailed knowledge about the cell which is the subject of the chapters that follow. Within this period, increasingly advanced methods of observation and of preparing cells for observation have greatly contributed to the overall progress that has been made. In all cases the refinement of techniques has been the outcome of collective enterprise; a catalog of names of all those involved would stretch to many pages.

The classical microscopists, from the seventeenth century down to the closing decades of the nineteenth, looked chiefly for two things: high magnification and crisp definition — freedom

Many of the main steps forward in biology and medicine followed the development of equipment that allowed researchers to penetrate the miniature world populated by cells and tissues. Even a simple

magnifying glass, revealing the unique whorls of a fingerprint, presents more detail than the unaided eye. The enlarged and stained section of human skin (center) is the view through a high-

power optical microscope. The scanning electron micrograph next to it shows the skin's surface. Selective staining is a powerful technique, used (right) to reveal the structure of membranes round the human brain.

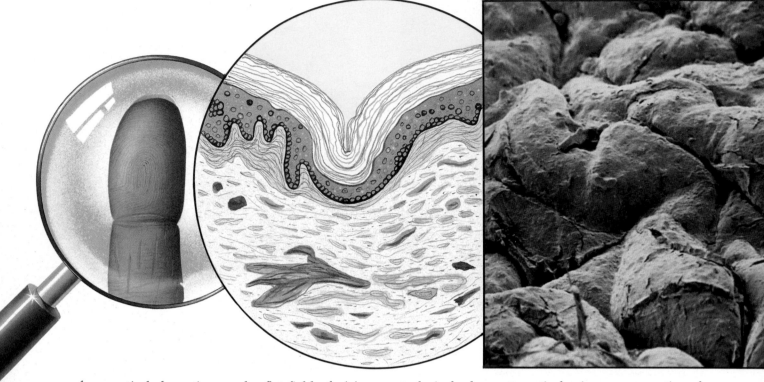

from optical aberrations and a flat field of vision. Measures to effect both were often in conflict, and it became apparent that magnification by itself was of little value without good resolution — after all, it was a simple business to project an image on a screen, producing enormous magnification without good resolution. Resolution is the minimum amount by which two adjacent points can be separated without losing their discrete appearance. The resolving power of a microscope depends on the lens closest to the object (the objective), the nature of medium between the object and objective, and the wavelength of light used.

In the late nineteenth century resolution was greatly increased by the microscopes devised by the German physicist and inventor Ernst Abbe and by the introduction of new techniques, such as oil immersion. Oil was placed between the sample and the objective lens, a technique that was effective in permitting better resolution. Cedar oil, for instance, was used as such a medium initially in England in 1870.

Following the invention of a host of other technical advances, optical microscopy continued to improve. In 1904 August Köhler showed that animal cells could be photographed by means of ultraviolet light; using this method he was able to observe nuclei dividing in the cells of salamander larvae. By decreasing the wavelength of the radiation used, ultraviolet microscopy doubled the maximum theoretical degree of resolution obtainable.

The advantages of ultraviolet microscopy may be extended by the application of fluorescence, often used as a powerful tool for the detection of very small quantities of material. Substances that do not naturally fluoresce can often be linked to fluorescent dyes (fluorochromes) to make them visible, a special example of ''staining'' as a procedure to produce contrast, and reveal objects not otherwise visible.

The effectiveness of ordinary light microscopes has also been increased by the development of techniques such as dark-field illumination, phase-contrast microscopy and interference microscopy. Dark-field illumination, as the name suggests, illuminates the object against a dark background.

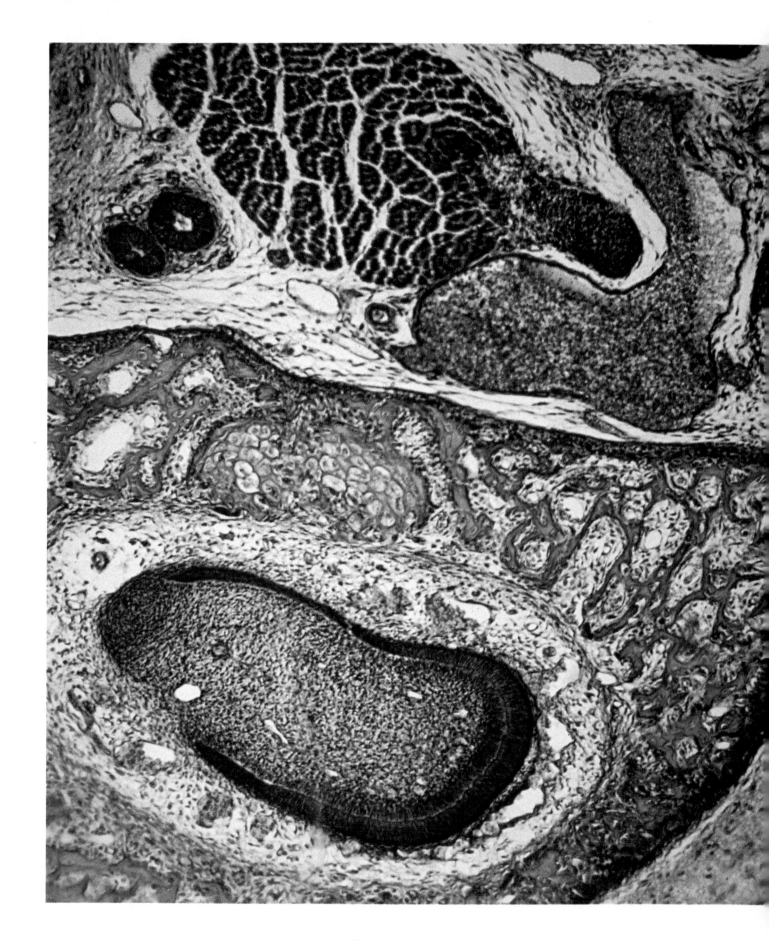

The optical microscope is limited to the degree of useful magnification it can offer. This limit depends on the ability of the objective to resolve fine detail, on the refractive index of the medium between the object and objective (air in the diagram below), on the illumination and on the wavelength of light used. With white light, air as the medium, and a good objective, magnification of about 1,000 times is the practical maximum.

The orange-yellow robes of the Buddhist monks have been stained with saffron. This dye was one of the first to be used for staining specimens for microscopy (by van Leeuwenhoek).

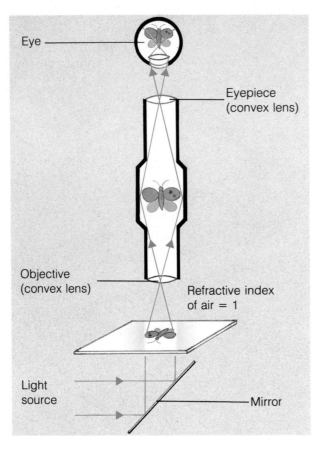

used. The result is low contrast, but such bright-field illumination, together with its complementary dark-field version, phase-contrast and interference techniques, all constitute methods of using optical means to increase the contrast in specimens without the need for staining.

Staining was a method known to microscopists as early as the beginning of the eighteenth century; van Leeuwenhoek, for example, used saffron to stain muscle fibers. A vast range of stains is now available to biologists for use both in histology (the study of the arrangement of cells in tissues) and in cytology (the study of the cells themselves). Some of the more commonly used stains include methylene blue, toluidine blue, safranin, eosin, and hematoxylin. The choice is large, and each stain has advantages and disadvantages. Some dyes are used only for contrast, and convey little other information. Such is the hematoxylin and eosin combination that is standard in many pathological laboratories. Other dyes can be used to learn about the chemical composition of cells. Thus the periodic acid Schiff (PAS) reaction reveals sugar-containing compounds in a cell; other stains can be used for protein, RNA or lipids.

Often the best way of studying a biological specimen is as a very thin slice or section. But a great difficulty in preparing such a specimen is getting a section of tissue which is undamaged in cutting. Slicing as finely as possible by hand — the method employed for centuries — inevitably resulted in too thick a sample. Biologists have been aided immeasurably by the development of the microtome, in which an extremely sharp blade cuts a piece of tissue embedded in wax or plastic at increments of 5 or 10 micrometers or even less. Microtomes were originally worked by hand but are now usually automatic.

The Electron Microscope

Ultimately the magnifying power of an optical microscope is limited by the wavelength of the radiation being used. Thus if the radiation is visible light, it is obviously impossible to use it to see anything smaller than a wavelength of visible light. In the 1930s several researchers — principally the electronics engineers Vladimir Zworykin and James Hillier working at RCA in the United States

This is ordinarily effected by light directed from the side; only scattered light enters the microscope lens, and the result is that the cell appears vividly lit on a black field. Phase-contrast and interference microscopy use the fact that when light passes through a living cell, the phase of each light wave is altered according to the relative density of the cellular structure and contents it passes through; an interference effect is thus established between waves of light and produces an image of greatly improved clarity.

All of this is in contrast to passing light through a dead cell that has been stained in order to increase visibility. In such cases the amplitude of the light waves is decreased at specific wavelengths, and a colored image of the cell is directly visible.

Occasionally, the technique of passing light directly through an unstained living cell is also

— hit on the idea of using a beam of electrons as an extremely short-wavelength type of radiation for microscopy. The wavelength depends on the voltage used to accelerate the electrons, and so high voltages produce high magnifications, up to 500 times better, or more, compared with those of the best optical microscope.

Working in a vacuum and with electromagnetic focusing devices instead of glass lenses (because electrons will not travel through air or glass), the new electron microscope opened up a whole new world to the cell biologist, revealing hitherto unknown structures at tremendous magnifications. Viruses, molecules and even individual atoms have been resolved by high-power electron microscopes. Images are generally viewed on a television-type screen, and recorded by letting the image-forming electrons expose a piece of film.

For transmission electron microscopy (TEM), in which the electron beam passes through the specimen, the specimen has to be extremely thin — up to 100 times thinner than ordinary wax sections. Ultramicrotomes, with glass or diamond knives, cut the required sections from specimens embedded in relatively hard methacrylate or epoxy resin. They are so thin that they display rainbow-colored interference patterns, like oil films on a puddle. To examine surfaces, a replica or mold of the specimen is made in plastic and coated with carbon (for strength) and a thin layer of metal (to improve contrast).

A better depth of field permitting a three-dimensional view of surfaces is obtained by scanning electron microscopy (SEM), which as its name suggests uses a narrow beam of electrons to scan the specimen. (The television-type picture is produced by amplifying secondary electrons emitted by the scanned specimen.) Usually the specimen is made electrically conducting by coating it with a very thin film of metal, applied by

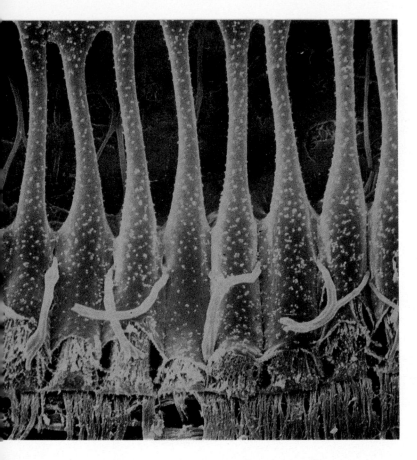

removal of ice in a vacuum. The technique prevents the distortion that occurs in tissue dried at ordinary temperatures; and when the time comes for the cells to be examined, they thaw without losing any of their former characteristics or even any enzymes they may have contained. Freeze-drying and critical-point drying are often used for preparing samples to be examined by a scanning electron microscope.

Other highly sophisticated methods of preparing materials for microscopical observation include microincineration, microdissection and micromanipulation. As early as 1833 it was demonstrated that leaves and other parts of plants could be oxidized by being burnt to ashes (calcined), leaving behind mineral deposits which still retained the general structural characteristics of the original material. In the twentieth century the technique has been modified to allow animal tissue to be calcined on slides, even such delicate tissues as nerve cells or cells of the retina of the eye.

Microdissection — either freehand or mechanical — is of particular value to embryologists, who need to be able to separate individual cells called blastomeres in developing eggs. Modern, refined methods use micromanipulators, instruments which operate needles and pipettes in three different planes under the microscope. Cells are held in place by mild suction on a pipette, and another can be employed to inject fluids or to remove, say, the cell nucleus. Specially elegant are procedures to remove selected portions of the membrane which remains stuck to a special pipette by a so-called Giga seal. Electrophysiological techniques permit the study of the permeability of these membrane patches to answer questions about the passage of ions and molecules through the membrane, questions vital to the understanding of how nerve cells function, for example. Another modern coup is the ability to inject live cells with reagents that permit the researcher to follow the behavior of natural substances inside the cell. Chemically isolated natural compounds can be labeled with a fluorescent group, injected into live cells, and used as a tracer to follow the behavior of the parent compound which it mimics using fluorescence microscopy (often enhanced by special video techniques).

evaporation from a hot filament in a vacuum. In some circumstances an antistatic spray may be used for this purpose.

Preparing Specimens for Study

In biology it is desirable that cells should be studied live but this is not always possible, especially with electron microscopes which have to operate in a vacuum. Often researchers are compelled to study cells preserved so that they retain as much as possible of their living appearance. Biologists therefore take great care in choosing a "fixative," or preservative. Ethyl alcohol was used in the seventeenth century, followed by various other compounds such as picric, chromic and acetic acids, potassium dichromate and mercuric chloride. Today the most commonly used general purpose fixatives contain aldehydes (formaldehyde or glutaraldehyde), to which osmium tetroxide is added for electron microscopy work.

A widely favored present-day technique is freeze cleaving, in which frozen samples are fractured in a vacuum and the broken surface "shadowed" by a thin layer of metal. This approach has allowed detailed studies of the inside of cell membranes. Another method is freeze-drying, in which tissue is frozen very quickly then dehydrated by the

Viruses are responsible for many diseases: the rotaviruses shown below, for example, cause gastroenteritis. They are so minute as to be seen only with the high magnifications of electron microscopy.

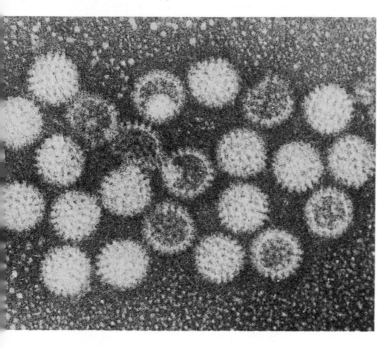

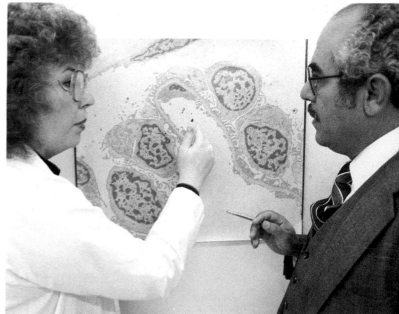

The electron microscope has made it possible to investigate the cell in previously undreamed of detail, enabling scientists to study aspects such as the effects of disease at a cellular level.

The construction and design of microscopes themselves have greatly improved in recent decades. One example is the increasing use of computers together with various types of microscope to enhance a picture received on a television screen. Investigative techniques have improved to include the use of sound waves (a beam of sound as opposed to particles) in the acoustic microscope, and the measurement of the absorption of high-frequency radio waves (providing a spectrum of nuclear magnetic resonance, NMR). NMR spectra are also used as diagnostic aids, particularly in the detection of tumors.

The biologist E. B. Wilson remarked in 1925 that the unfolding of cell theory was a turning point in the history of biology, one of the commanding intellectual achievements of nineteenth-century science. The key to every problem in biology, he suggested, was to be found in the cell, because every living thing had once been a single cell. And in an entirely practical sense the key to the study of cells has been the development of microscopy. As Sir Ralph Peters once put it, "the cell is the most important invention of Nature and must be the continuous wonder of any thinking person."

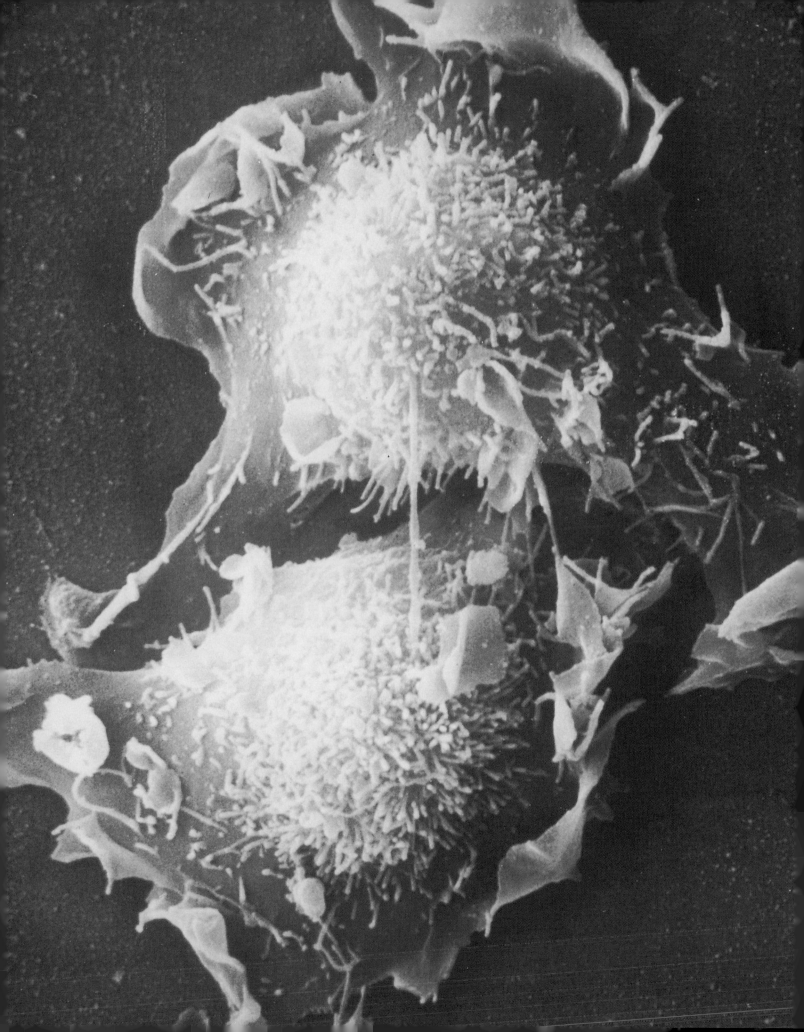

Chapter 2

Marvels of a Miniature World

The cell is the fundamental unit of life. Nothing smaller than it is truly living — that is, capable of independent existence (within its environment) and of self-replication. All cells require various key components and structures: a cell membrane to hold them together, enzymes — biological catalysts — to bring about biochemical processes, internal membranes to encapsulate "packets" of chemicals, and genetic material to provide the information for producing the components and for their own replication.

Cells are divided into two main types by the way the genetic material is organized inside them; they are known as procaryotic and eucaryotic cells. In simple, procaryotic cells, the genetic material and the enzymes required for energy production, cell growth and division are contained in the jellylike cytoplasm which is surrounded by a boundary or plasma membrane. Although the mixture of chemicals within is highly complex, the cells have only a simple internal organization. Procaryotic cells do not have a nucleus bounded by a membrane and their DNA (deoxyribonucleic acid) is attached to the plasma membrane.

Eucaryotic cells, on the other hand, are much larger. They have a separate membrane-bound nucleus which contains the genetic material. In general, eucaryotic cells are larger and much more complex than procaryotes, which are mostly single-celled organisms such as bacteria and cyanobacteria (blue-green algae). The fluid filling of eucaryotic cells is divided into the nucleoplasm, situated inside the nuclear membrane, and the cytoplasm placed outside the nuclear membrane.

Many eucaryotic cells are highly specialized and able to perform different functions which are carried out by means of intracellular compartments known as organelles. Most of the organelles are contained in the cytoplasm.

Each organelle interacts with other parts of the cell, importing essential chemicals from outside

Cells multiply by splitting in two. Here at the moment of division, two new "daughter" cells — still attached to each other by a slender strand — are formed from a single parent in the miraculous process of mitosis.

31

The most significant, categorizing feature of cells is the presence or absence of a nucleus. Paramecium (below), a single-celled organism, is a eucaryotic cell — that is, one having at least one nucleus. Nearly all human body cells are eucaryotic, as are those of all living organisms except for bacteria and cyanobacteria (blue-green algae). These exceptions are procaryotes, cells with no membrane-bound nucleus.

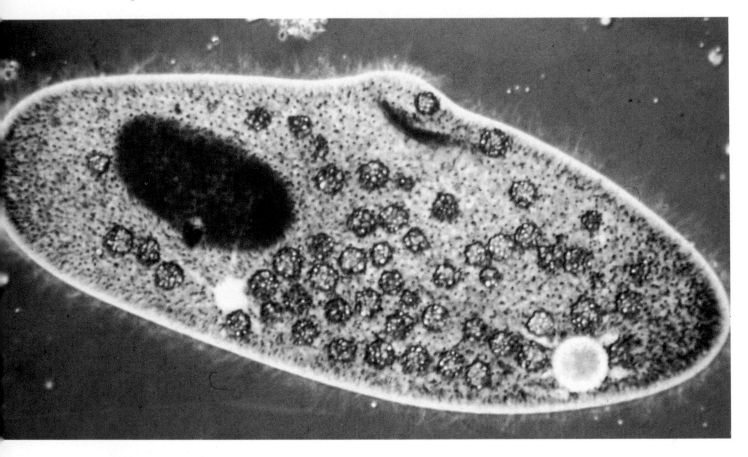

themselves, and exporting the cell's products. Organelles help the economy of the cell in two major ways: they concentrate the components needed for one particular job, such as protein synthesis, so that the reactions occur efficiently. They also allow certain functions to be isolated from the rest of the cell so as to protect it from its own potentially dangerous elements. For instance, oxidizing enzymes would destroy the cell that produced them if they were not isolated in some way.

Lewis Thomas, the former President of the Memorial Sloan-Kettering Cancer Center, drew an analogy between the cell and the planet Earth when he compared cellular organization with the way in which oceans, landmasses and winds all interact with each other to produce local climate just as the organelles within the cell function to produce specific intracellular effects. Another analogy is the comparison of a cell with a factory. Chemicals are taken up by the cell and processed in one of several internal production lines. The manufactured products can then be used to build more cellular material or may be exported into the environment.

The Cell Boundary

Every cell, whether eucaryotic or procaryotic, has an outer plasma membrane which acts as a boundary between the cell contents and the environment outside. Substances must, however, be able to pass in and out of the cell. Glucose, potassium ions, and growth-regulating substances such as some hormones bind to and are taken up by the cell, other substances such as enzymes are secreted, and waste products have to be passed out.

The plasma membrane is extremely thin — about 0.00001 millimeters thick — and its structure is essentially the same in all living organisms. It is

Medieval European villages were maintained by peasants who worked continually at mixed agriculture, tilling, sowing, reaping, herding animals and cultivating fruit. Some produce was for their own use and some for trade, and all was regulated by the lord of the manor. In a similar way the various structures in a cell work to produce enough protein and energy for the cell itself and for export to the cell's exterior. These activities are moderated by the nucleus, the administrative center of the cell. In both cases, the benefits are mutual for the workers and the controller — neither could survive on its own without the other.

composed of a double layer of phospholipids. The ends of the phospholipid molecules are chemically different from each other — one end is water-loving (hydrophilic) and the other is water-hating (hydrophobic). The molecules look like old-fashioned wooden clothespins, with the hydrophilic end at the head and the hydrophobic portion forming the two legs. The lipids line up in two layers (a bilayer) with the hydrophobic tails of one layer pointing towards those of the other on the inside. The hydrophilic ends, which can interact with water, point to the outside. The proteins that form a major part of the cell membrane are incorporated in the bilayer.

When scientists looked at the plasma membranes with an electron microscope, they saw them as three-layered structures, with dark outer layers sandwiching a clear central zone, and some thought that the membrane proteins probably floated in the upper and lower surfaces of the lipid layers. But this "trilaminar" model is probably incorrect because it does not fully account for the properties of membrane proteins. Although some are at the membrane surface, others pass through the membrane leaving opposite ends exposed on each side.

The presently accepted theory of membrane structure is the "fluid mosaic" model. This suggests that the membrane proteins are like icebergs floating in a shallow sea — the icebergs are the membrane proteins and the sea is the lipid layer.

The plasma membrane is the cell's interface with the world around it. All interactions between the environment and the cytoplasm inside the cell must involve some part of the membrane. These interactions are largely controlled by the membrane proteins, which may join with molecules floating in solution outside the cell and cause them to be taken inside. The proteins may also bind the cell to its neighbors so that groups of cells are formed. This is

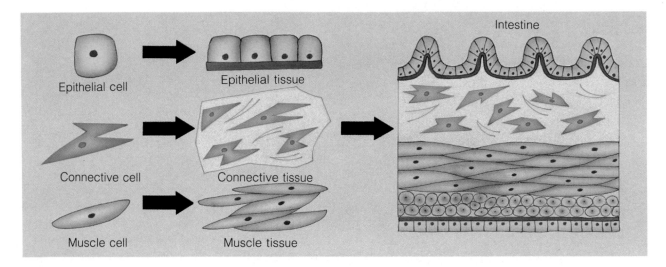

thought to be a first step in the origin of multicellular higher forms of life.

Movement of substances in and out of cells is either a passive process or an active, energy-consuming one. Passive transport can occur by simple diffusion down a concentration gradient — from a region where a substance is present in high concentration to a region where it is present in low concentration (until both regions equalize). In this way, a few types of small molecules such as those of oxygen and carbon dioxide can pass freely across the membrane.

Other molecules such as glucose, or sodium and potassium ions, however, cannot cross the membrane on their own by simple diffusion and need selective transport proteins or special channels to allow them in and out of the cell.

In active transport, the cell has to do work to bring particles in or push them out. For example, virtually all animal cells have to maintain a high concentration of potassium and a low concentration of sodium inside the cell. But outside the cell the concentrations are reversed, with the result that a mechanism termed a sodium-potassium pump is needed to maintain the correct balance inside the cell. The energy needed for active transport is derived from molecules of a high energy phosphate known as ATP (adenosine triphosphate). When one ATP molecule is broken down biochemically, energy is effectively released which allows three sodium ions to be pumped out of the cell through the membrane and two potassium ions to be taken in. Other substances are transported through the cell membrane in a similar way.

Cells also use another mechanism to move molecules into the cytoplasm — endocytosis — which includes the processes known as phagocytosis and pinocytosis. Phagocytosis is the taking in of bacteria or other foreign bodies. Cells conduct such an operation by flowing around the particles and engulfing them. The contents of the resulting sacs or vacuoles are digested, and the waste is passed out in a process called exocytosis.

Pinocytosis involves the ingestion of fluid by the inward movement of part of the membrane. An area bulges into the cell like a pocket, and fluid is trapped as the membrane closes around it. The encapsulated fluid or vesicle moves through the cell cytoplasm and fuses with an organelle called a lysosome, in which the contents are broken down by enzymes. The vesicle membrane then returns to the cell membrane.

Another form of endocytosis uses specialized regions of the plasma membrane called "coated pits." These involve a unique protein, clathrin. Unlike random endocytosis, the coated pits ingest the selected molecules along with specific receptor proteins in the membrane. The receptors for the selected molecules of low density lipoprotein or LDL (a major source of cholesterol, itself a building block of steroids and other substances) are scattered at random over the cell membrane, but when they bind an LDL molecule, they move around until they reach a clathrin-rich region. The coated pit develops as clathrin forms a structure like a pocket into the cell, creating a vesicle which contains the receptor-bound material. The vesicle fuses with a lysosome, the LDL is digested, and the receptors return to the cell surface.

Cell membranes have receptors for many other substances; the hormone insulin for example attaches to the cell by this mechanism. Some receptors occur only on a few types of specialized cells. A good example of these are B-lymphocytes, white blood cells formed in lymph glands, which are part of the body's defense system. A B-lymphocyte has an antibody in its membrane which can bind a specific antigen on a "foreign" protein which may be a disease-causing organism or a toxin. In simple terms, when the receptor binds the antigen, the lymphocytes are stimulated to multiply and produce antibodies to neutralize it. Other lymphocytes with receptors for different antigens are not activated, and hence the antibody response is specific for the antigen which induces it.

The Cytoplasm

The jellylike cytoplasm within a cell is made up of protein-containing cytosol, water, minerals, organelles that carry out specific functions, and various inclusions. Organelles are permanent structures and are vital to the continuing life of the cell. Inclusions are particles of material, such as fat globules and waste products, that are stored in the cell. In a typical human cell the volumes occupied by cytosol and organelles are roughly equal.

The cytoplasm is the site of most of the "intermediary metabolism" of the cell, in which some molecules are broken down to produce energy and other molecules are built up. Within the cytoplasm is the cytoskeleton, networks of protein filaments that act like flexible scaffolding to maintain the cell's shape and allow it to move. The shapes and positions of the cytosol, organelles and cytoskeleton sometimes change, which makes them difficult to study. This problem is further exaggerated when they are artificially "locked" by chemical fixation (so that they can be studied under an electron microscope) into an arrangement which may not be typical.

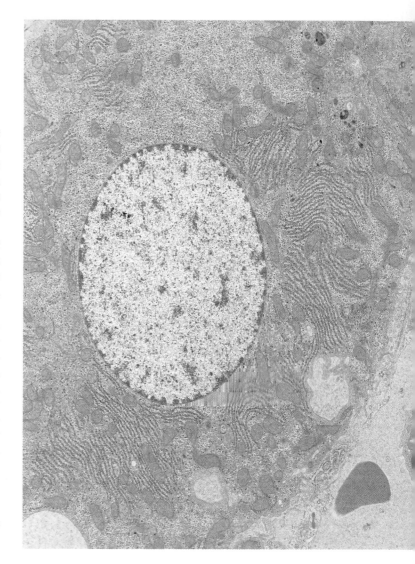

The Nucleus

The "reference library" of the cell is the nucleus. It contains nucleoplasm and is bounded by the nuclear envelope (consisting of two membranes) which is pitted with holes, or pores, connecting the nucleoplasm with the cytoplasm. The nucleoplasm holds a "file" of information — genetic material in the form of DNA molecules. The DNA is packaged into a number of chromosomes, which consist of long strands of DNA packed tightly into coils with proteins called histones. Eight histones make up a nucleosome. The DNA is wound twice around each set of eight histones like thread round a spool. Numerous nucleosomes serially wound with the same thread combine together to form a chromosome. The combination of DNA and histone proteins is known as chromatin.

Normally, chromosomes occur in pairs, but before cell division in the process of mitosis each chromosome is duplicated to form two identical

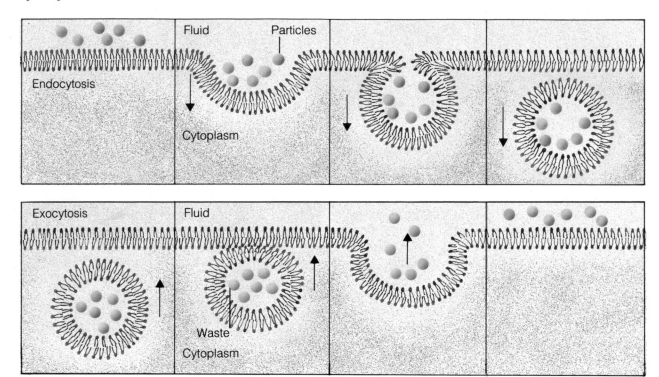

chromatids held together by a centromere. A specialized series of rope-like protein strands or spindle fibers, which are attached to the chromatids at the centromere, pull the chromatids apart in opposite directions. When a cell divides, each daughter cell receives one of the two copies of the parent cell's chromosomes. The nucleus also contains a nucleolus, a structure which controls the synthesis of some of the cell's RNA (ribonucleic acid). Some types of RNA are used to make ribosomes, which are the "machine tools" of protein synthesis existing singly or in small groups within the cytoplasm (polyribosomes), or attached to the endoplasmic reticulum.

There is a continual two-way traffic of molecules in and out of the nucleus through the nuclear membrane. All the proteins required for nuclear structure and function are imported into it from their sites of synthesis in the cytoplasm, and all the RNA required for protein synthesis is exported. The nuclear membrane also has sites on its inner surface to which the cell's chromosomes bind. This ensures that the strands of DNA do not become entangled with each other.

The Ribosome Factory

Proteins perform most of the significant chemical reactions that occur in the cell. They are also important in maintaining its structure. They are the cell's tool kit and framework.

Proteins are long strings of amino acids attached to one another like beads in a necklace. Different proteins have different sequences of amino acids which are determined, or coded, by the DNA. In protein synthesis, an RNA copy of the DNA of a gene is transported to the cytoplasm, where ribosomes, other RNAs and enzymes all come together to translate the RNA structure into a specific amino acid sequence, or protein. At any one time in a particular cell, RNA copies of most genes are not made; only the genes that are needed for the cell's functions are translated by that particular cell. Some proteins are made in one type of cell and not others because the appropriate gene activators are switched on only in that type of cell.

Protein synthesis occurs through the interaction of three kinds of RNA molecules. Messenger RNA (mRNA) provides the coded message for the amino acid sequence; transfer RNA (tRNA) carries the

DNA

mRNA

Polypeptide chain

Amino acid

tRNA

mRNA

Ribosome

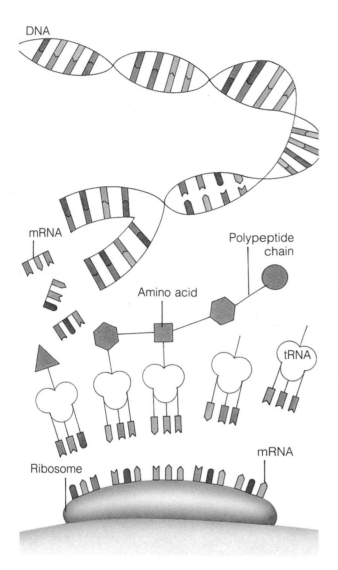

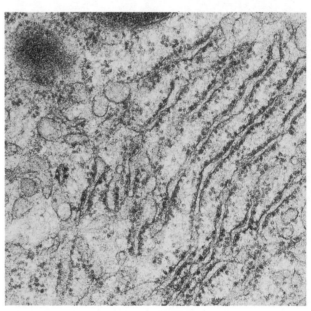

When DNA unzips during protein synthesis, messenger RNA (mRNA) is made as a copy of the unzipped part and breaks away. It then attaches itself to a ribosome. Transfer RNA (tRNA) molecules with the same pattern arrive and bind, each carrying an amino acid. Once they slot into the mRNA, the amino acids separate from the tRNA and link to each other to make a polypeptide chain, which forms a protein.

amino acids being assembled into proteins and has a small decoding region that interacts with the message; ribosomal RNA is bound into particles that act like a sewing machine, attaching the amino acids to one another while rolling along the message.

Ribosomes are cell organelles composed of RNA and protein. They appear to have three slots or grooves on their surface, one which holds the mRNA and two which hold the tRNAs while enzymes bound to the ribosome attach amino acids — in the correct sequence — to the growing protein. Ribosomes are large enough to see with an electron microscope and it appears that ribosomes and protein synthesis can occur either freely in the cytoplasm when the protein is used in the cell only, or on the surface of a membrane-bound set of channels known as the endoplasmic reticulum, when the protein synthesized is exported from the cell to other parts of the body.

The Endoplasmic Reticulum

Cells have a complex of sacs enclosed by a folded sheet of membrane, the endoplasmic reticulum, within their cytoplasm. One part of the sheet is joined to the outer surface of the nuclear membrane. If it is imagined as a simple, flat membrane folded into a closed oval shape in the cytoplasm,

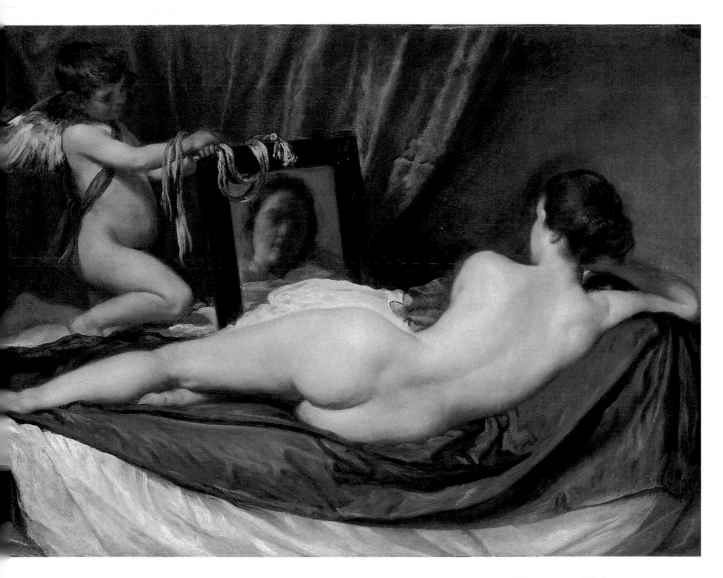

The artistry of Velasquez captures the smooth translucency of a woman's skin. The outer surface of human skin is made up of the same substance as hair and nails — the intermediate filament protein keratin. This accumulates in epithelial cells and is synthesized by the mechanism diagrammatized on the opposite page.

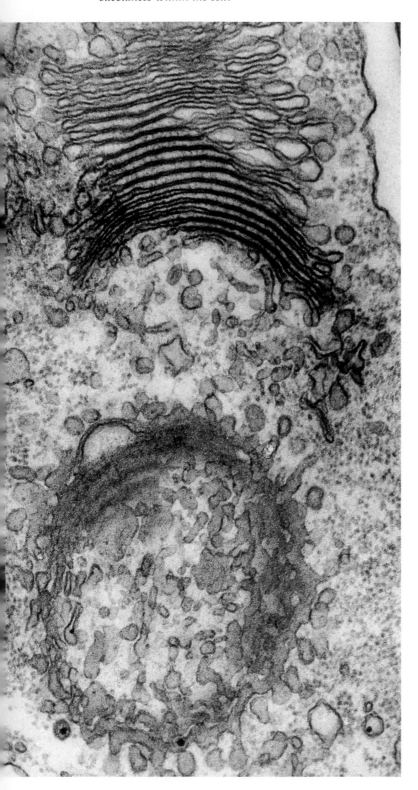

The Golgi apparatus, seen here in longitudinal section (top) and in transverse section (bottom), is concerned with the synthesis and transport of various organic substances within the cell.

then it has two surfaces or sides. One surface (the outer one) is in contact with the rest of the cytoplasm and the other surface is "inside" it, totally separated from the cytoplasm so the inner surface is confluent with the inner lining of the nuclear envelope. The internal space is the cisternal space or lumen. The membrane controls the passage of molecules from the cytoplasm to the cisternal space. This is important because proteins are assembled on the cytoplasmic side, and must cross into the cisternal space if they are destined for export. It also ensures that potentially dangerous proteins such as digestive enzymes are kept separate from the rest of the cell as they are synthesized.

The endoplasmic reticulum has two functions. It has a major role in synthesizing macromolecules such as proteins, lipids and complex carbohydrates that make up some of the other organelles in the cell; and it separates recently-made molecules destined for the cytoplasm from those intended for transport to other sites.

Electron microscopy shows that there are two distinct types of endoplasmic reticulum, rough and smooth. The rough or granular form has ribosomes attached to its outer surface and is a site of protein synthesis. The smooth type has no ribosomes and is, instead, a site of lipid synthesis. Lipids are required for the growth of cell membrane and for the membranes of organelles within the cell. The two kinds of endoplasmic reticulum, although part of the same internal sheet of membrane, are organized into different shapes. The rough sort encloses thin, flat spaces, whereas the smooth one is composed mostly of fine hollow tubes.

The proportion of rough to smooth endoplasmic reticulum varies depending on the function of the cell. Large amounts of the rough type occur in cells such as B-lymphocytes (in which antibodies are formed). Such cells are engaged in making large quantities of protein for export. The smooth form is prominent in cells that specialize in lipid metabolism, such as steroid hormone production.

Rough endoplasmic reticulum has specialized proteins in its membrane that bind ribosomes to its cytoplasm side. The growing protein is pushed through the membrane so that it extends into the inner cisternal space. Here sugar molecules are added to the protein. This is an important part of their final processing, particularly because the sugar chains may act like a kind of "zip code" to direct the protein to the correct destination for its final function.

All proteins destined for export start with a short sequence of joined amino acids, called a leader sequence. This sequence is the first bit of the protein to be made and is captured along with its RNA and attached ribosomes, by a specific protein on the endoplasmic reticulum. The leader sequence is often removed as part of the final processing of the protein. When a protein is complete it may be removed and pass into the cisternal space if it is intended for export, or it may be retained in the membrane if it is eventually to be an integral membrane protein.

Proteins and membranes made by the endoplasmic reticulum must be transported around the inside of the cell to reach the organelle or surface membrane that is their final destination. This is done by a type of vesicle formation similar to endocytosis — it is as if the newly-formed substances were being carried in protective plastic bags. The endoplasmic membrane forms "buds" which are pinched off, move through the cytoplasm and fuse with the membrane of their destination. If that is the cell surface, then the proteins contained may be released to the outside. If the target is an organelle within the cell, then the vesicle plus the attached proteins become part of the organelle's membrane, and proteins enter the target's internal space.

The Golgi Apparatus

Many of the vesicles produced by the endoplasmic reticulum are destined for the Golgi apparatus, a complex stack of interconnecting membranes and spaces located near the cell nucleus. The complexity of the Golgi apparatus makes it hard to isolate and study, so relatively little is known of its exact function. It seems to play an important part in processing and packaging newly-synthesized proteins into the correct final configurations. Once final processing is complete, proteins are removed from the Golgi apparatus and are moved to their final destination in vesicles. The membrane-bound vesicles fuse with the cell's plasma membrane and

The legendary figure of German folk literature Struwwelpeter, or Shock-Headed Peter, was renowned for, among other things, his extra-long fingernails, which he never cut. The keratin of which nails are made derives from protein filaments within the multiplying cells at the nail bed.

*Looking like an architect's model
of a futuristic building complex,
computer graphics display the natural
design of microtubules in the axoneme
core of a cilium, one of the "hairs"
that sprout from various cells.*

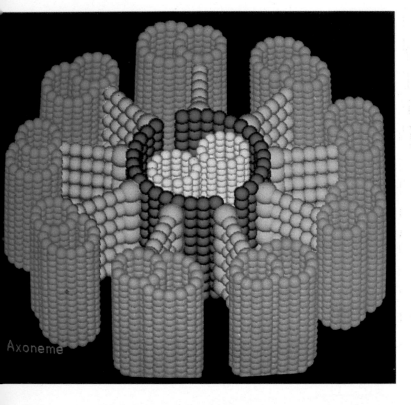

Axoneme

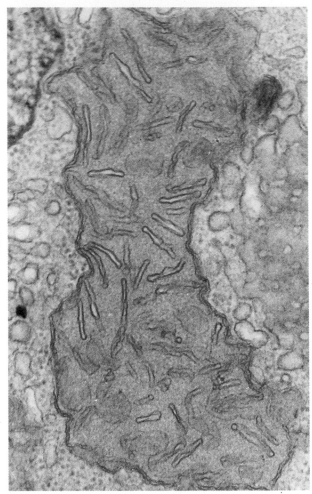

release their contents to the exterior using the process of exocytosis. In cells whose main function is secretory, the Golgi apparatus is well developed and the plasma membrane therefore has to be able to incorporate large quantities of intracellular membrane.

Two organelles, lysosomes and peroxisomes, are manufactured by the endoplasmic reticulum and the Golgi apparatus. Lysosomes are the cell's major digestive organelle and peroxisomes are important in detoxifying certain substances such as ethanol (alcohol) so that they do not poison the cell. Because both digestion and detoxification involve potentially dangerous enzymes or oxygen metabolites, they must be kept separate from the rest of the cytoplasm. Oxygen molecules, for example, can be broken down into "free radicals," which are single oxygen atoms that readily bind to proteins and lipids and destroy the normal structure of their recipient.

Lysosomes are vesicles, containing enzymes, that bud from the Golgi apparatus but which begin in the rough endoplasmic reticulum. They vary greatly in shape and size because they fuse with other vesicles to carry out their functions. A wide range of materials may be ingested by cells, from proteins to whole bacteria. Small primary lysosomes fuse with these ingested vacuoles and release enzymes which break down the contents into small molecules that can be used in the cell's metabolism. Occasionally a damaged organelle such as a mitochondrion also needs recycling by lysosomal digestion. This function of lysosomes, termed autophagy, results in breakdown products which are either reused by the cell or accumulate in the cytoplasm.

Peroxisomes are bounded by smooth membrane and are especially concerned with using oxygen in catabolic processes, which involve the breakdown of organic molecules into simpler ones. These processes range from "digesting" bacteria to breaking down drugs and alcohol. A recent hypothesis implicates these organelles in cellular aging, suggesting that they occasionally leak dangerous oxygen metabolites that damage the cell's contents and may even kill the cell.

Seen closeup (left) a mitochondrion consists of a smooth outer wall with an inner folded membrane; the folds, or cristae, create a large surface area on which the energy-generating reactions can take place.

Workers bring together various materials and assemble them to an architect's plan. In a similar way, various organic materials are brought together to build cells according to a genetic plan.

The Cell's Power-houses

Most eucaryotic cells (ones with nuclei) have organelles specialized for energy production — the mitochondria, which generate energy from sugars and fatty acids. These structures have a double membrane (rather than the single membrane of endoplasmic reticulum or the Golgi apparatus), the inner layer of which is thrown into complex folds that increase its surface area. Specialized enzymes that trap energy from the breakdown of sugar are embedded in the inner layer.

The number of mitochondria varies enormously depending on the cell type and may constitute 20 percent of a cell, such as a liver cell, that has a high energy requirement. In many cells the mitochondria are located next to a fat droplet in the cytoplasm. This is useful because the fatty acids which can be obtained from the droplet are a major source of the energy produced by the mitochondria.

Mitochondria have their own small circular DNA molecules and ribosomes that resemble those of procaryotic (non-nucleated) cells, and can synthesize their own membranes and at least some of their own protein, although many enzymes are imported from the cytoplasm. This has led to the suggestion that mitochondria might have evolved from bacteria that once developed a close relationship with primitive eucaryotic cells, and then lost the capacity to live outside the cell.

Mitochondria are self-replicating and so are inherited differently from all the other components of the cell (which are controlled by the genetic material in the nucleus). When a cell divides, all the newly-made molecules and organelles are created according to the genetic message in the nucleus, except for those in the mitochondria. These structures employ synthetic processes similar to those in the rest of the cell but use their own DNA and ribosomes. They divide by simply splitting in two. When a cell divides the mitochondria are randomly assorted into the cytoplasm of the two daughter cells.

During sexual reproduction most of the mitochondria in the fertilized egg cell are derived from the female sex cell and not from the sperm. This is unlike nuclear DNA which is equally derived from both parents, and means that many, and sometimes all the mitochondria in any one organism are derived from its mother. If this symbiotic hypothesis for the origins of mitochondria is correct, then independent inheritance of bacterialike organelles suggests that mitochondria have really been the most successful "life-forms" during evolution; they now live inside all living creatures and direct their own metabolism. Fortunately, their near monopoly of cellular energy production acts for the benefit of all the other parts of the cell.

Microfilaments and Microtubules

The distinct shapes of cells and their ability to move are determined by a complex network of proteins that form the cytoskeleton. This is organized into actin-containing filaments (microfilaments), microtubules and intermediate filaments. The microfilaments and microtubules perform the same supportive functions as tent poles, and similarly can be put up and taken down rapidly. The intermediate filaments are rather like the girders inside a building because they tend to be permanent, composed of stable, fibrous proteins. Although the elements interact to control the cell's shape and ability to move, they each make a unique contribution to the final arrangement.

Resembling bundles of threadlike structures, most actin filaments are concentrated just below the plasma membrane, where they have two important functions. They can support small projections from the cell surface such as some microvilli, which are important for absorption because they increase the cell's surface area. Many specialized cells have surface projections that are crucial to their activities. Most of these projections have an actin core organized by a specific accessory protein used by that particular cell type. They include hairlike cilia and whiplike flagella.

Actin filaments also form "micromuscles" when associated with myosin molecules. These complexes can produce contraction and movement by a rowing action of the myosin molecules along the actin fiber which is anchored to the plasma membrane. Actin and myosin filaments form the basis of contraction in smooth muscle. More highly organized sets of actin filaments also form the

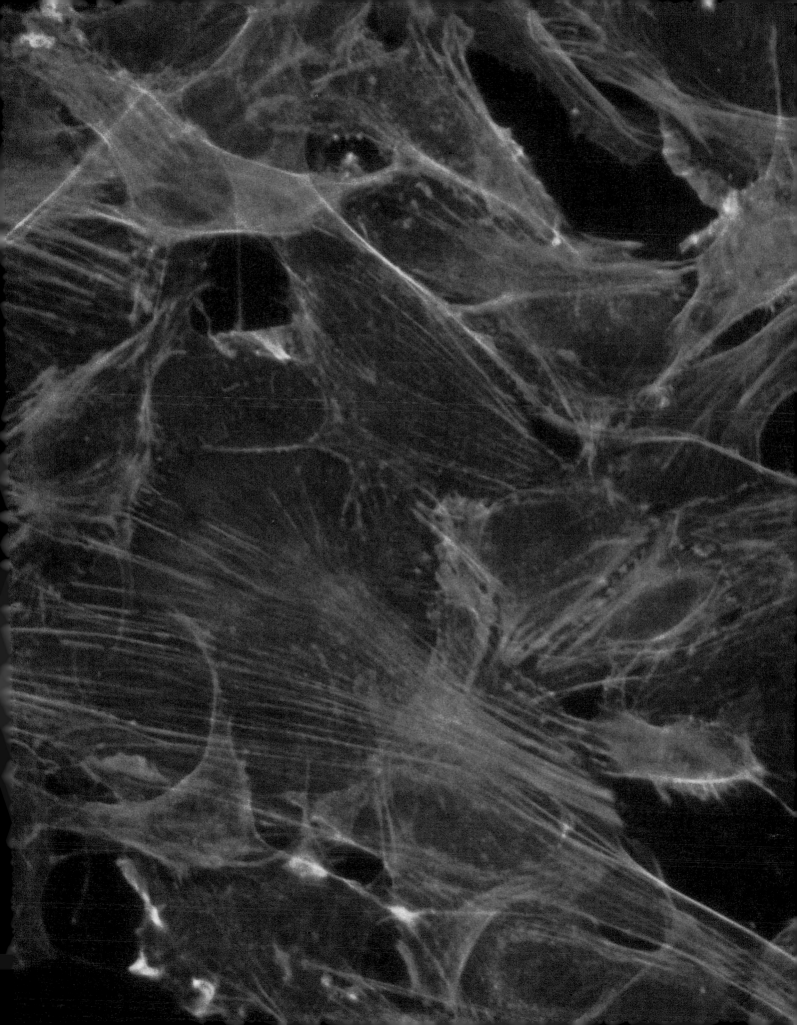

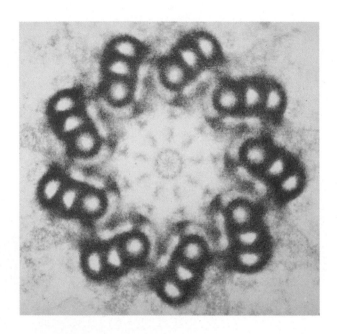

A centriole is a cell substructure consisting of microtubules which have a characteristic spatial geometry of nine sets of three tubes arranged to form a cylinder (here shown in cross-section). Each cell has two centrioles, which probably act as nucleation centers for the formation of microtubules for location elsewhere in the cytoplasm. The centrioles move apart to the poles of the cell in the preliminary stage (prophase) of mitotic cell division.

contractile elements of heart muscle and skeletal muscle for example; they also facilitate a variety of movement in cells.

Microtubules may determine the overall configuration of a cell's cytoskeleton. They are made up of subunits of the protein tubulin and seem to have their origin in and depend on the cell center, although they occur more densely around the inner edges of the cell. They are involved in the movement of chromosomes during cell division, the spindle fibers of the mitotic spindle being made of microtubules. The distribution of organelles in the cytoplasm and the activity of the plasma membrane are also associated with microtubules. They form a skeleton down the center of cilia and flagella (projections of the cell membrane).

The distribution of intermediate filaments, long ropelike strands that may act to hold the cell in shape, also is determined by microtubules. Keratin, which forms hair, nails and the outer surface of the skin, is an intermediate filament protein that accumulates in certain epithelial cells. When such cells die they leave behind their tough, crosslinked fibrous skeleton as hair fibers or fingernails.

The cell contains an exceedingly complex set of interconnected parts. Some of these are well understood because they can be seen using the techniques of modern microscopy and are sufficiently stable to be isolated. Other parts are best seen only in intact cells and may be destroyed by the very processes used to study them, which makes their fine structure difficult to determine and knowledge of their operation difficult to obtain.

The study of cells, their structures and functions continues to be an area of intensive research. Techniques continue to be devised and improved, and gradually the cell is being made to yield more and more of its secrets. With better understanding, scientists can apply the new knowledge in a wide variety of ways — particularly to plant and animal genetics, thus creating new strains of crops and farm animals, and to the study of diseases and methods of treating and preventing them. Eventually the techniques of genetic engineering may make it possible to construct, in a test tube, a tailor-made cell which could perform any desired function — a living factory for making useful biochemicals.

Long tresses of hair are admired as things of beauty. Soft and gleaming, often curling, but always growing, they seem to have life. It is hard to believe that they are no less than dead cells and consist of intermediate filaments — all that remains of the cells' structure.

Cells need digestive organs to process
and eventually eliminate waste
products. This is the function of
lysosomes (the large central
structure), which use enzymes to
bring about the chemical breakdown.

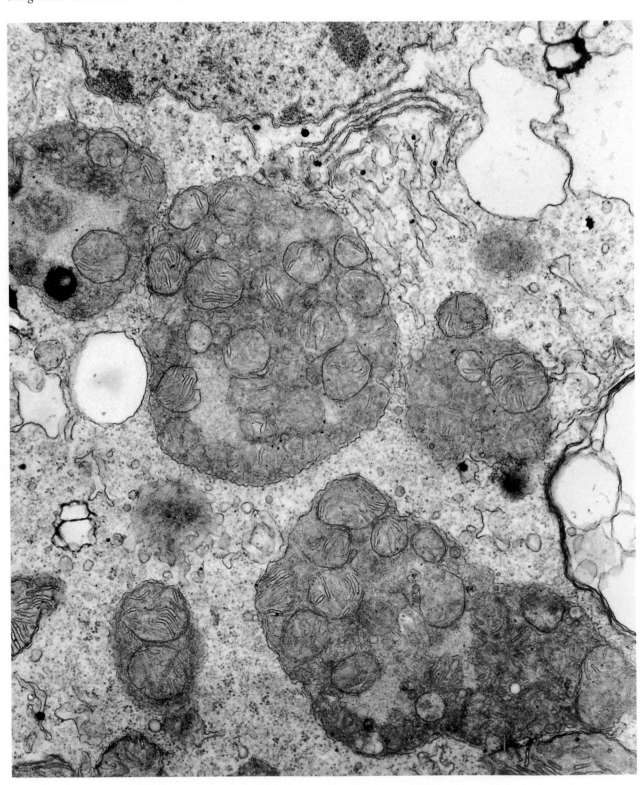

Most cells in the human body are roughly spherical. In many the outer surface is covered with microscopic hairs, spikes or other projections that resemble the fleshy tentacles of a sea anemone.

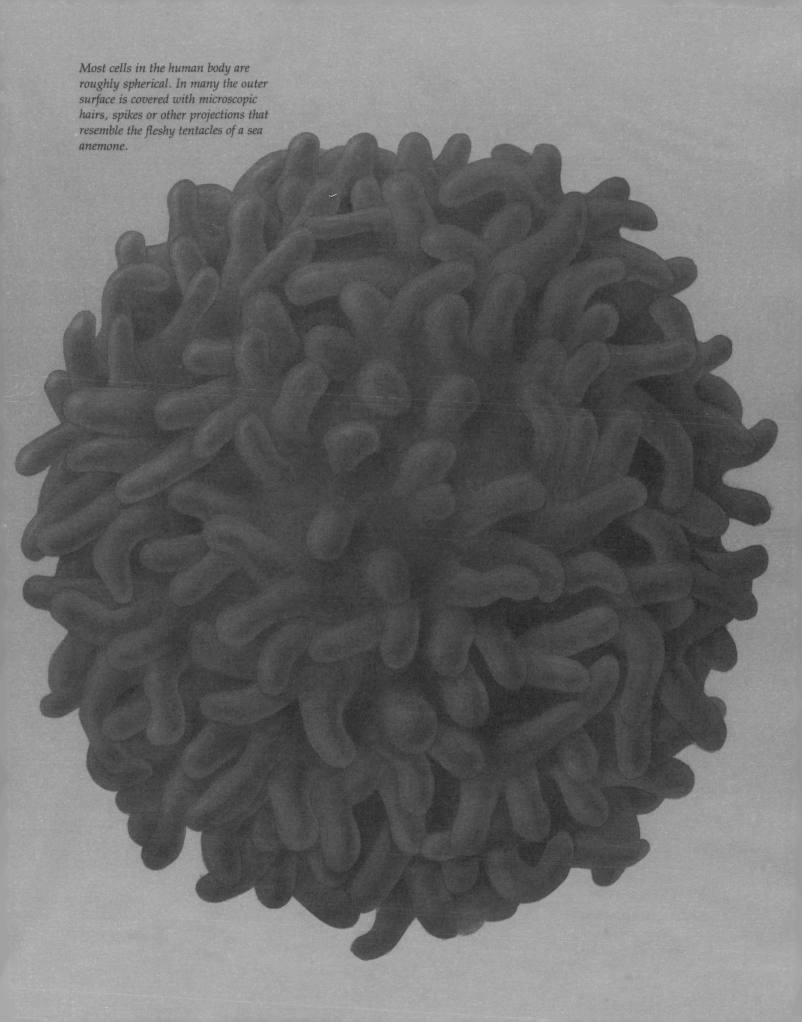

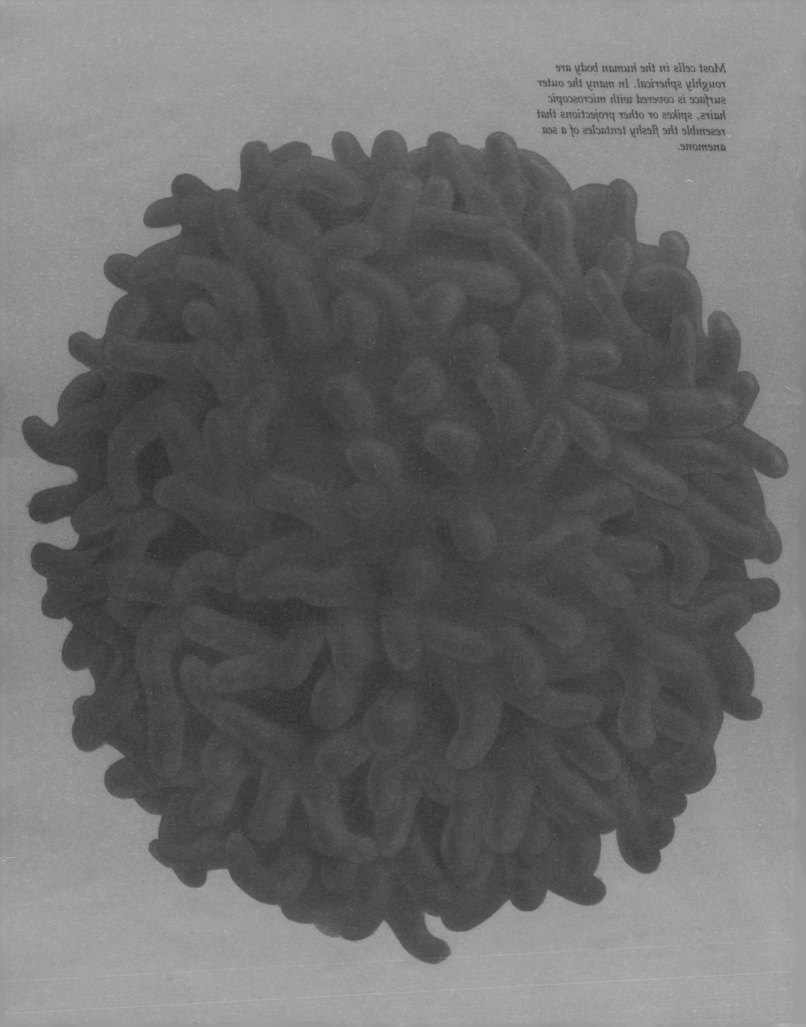

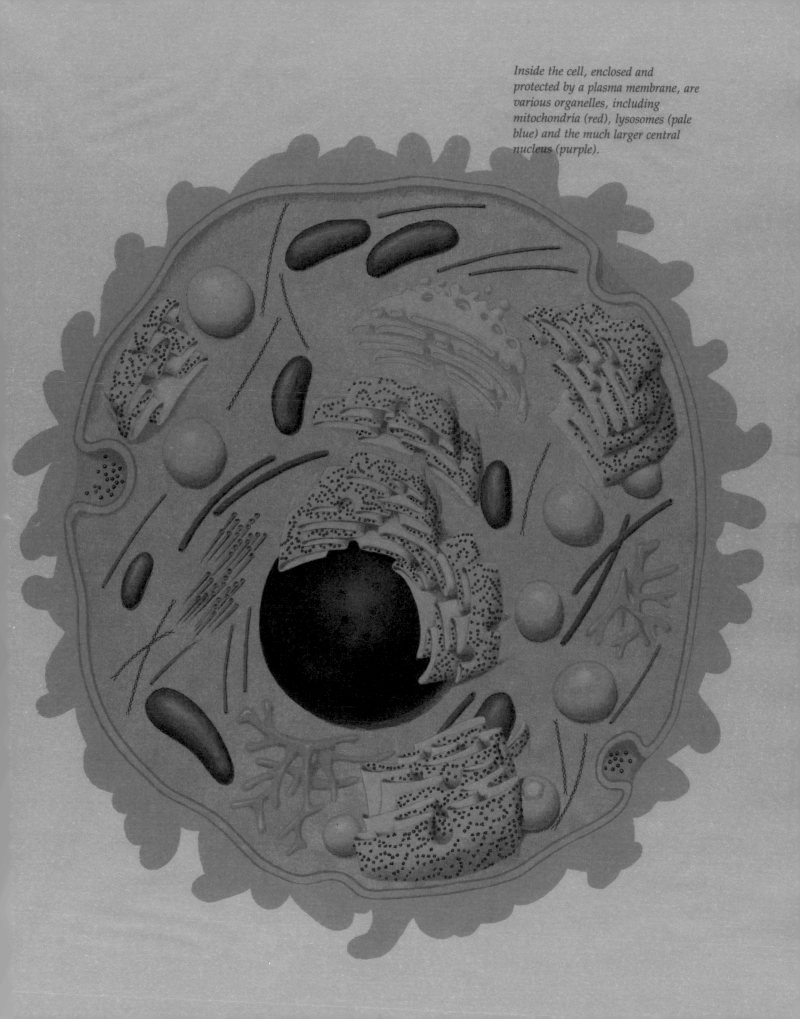

Inside the cell, enclosed and protected by a plasma membrane, are various organelles, including mitochondria (red), lysosomes (pale blue) and the much larger central nucleus (purple).

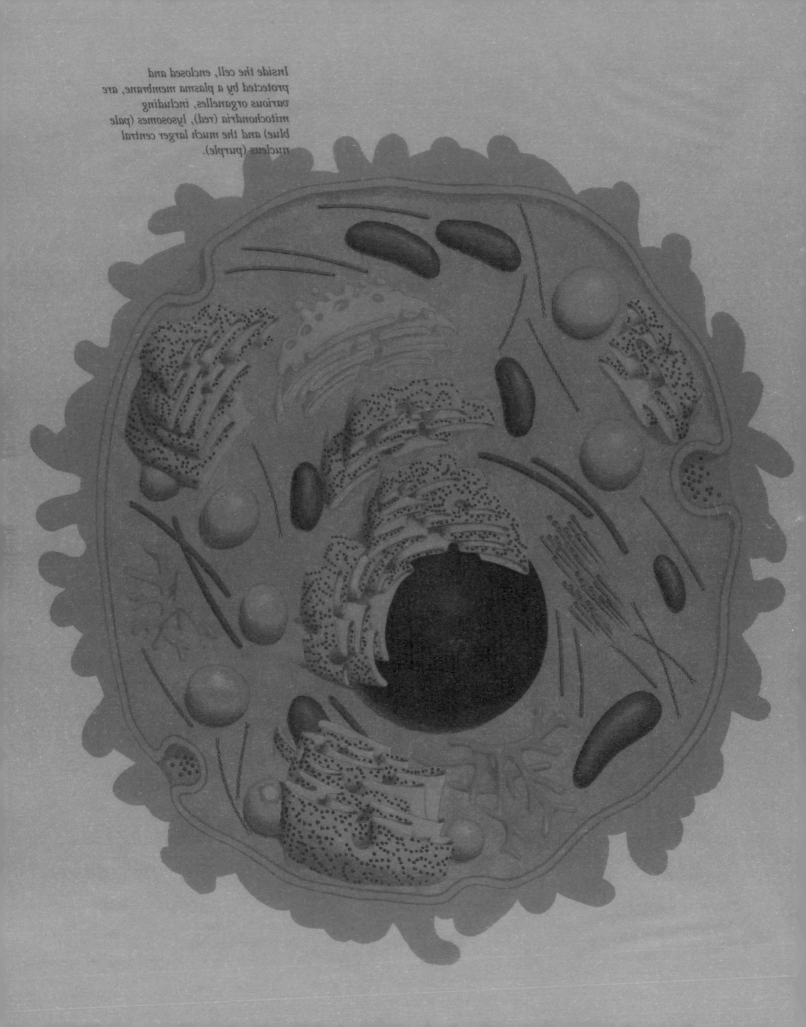

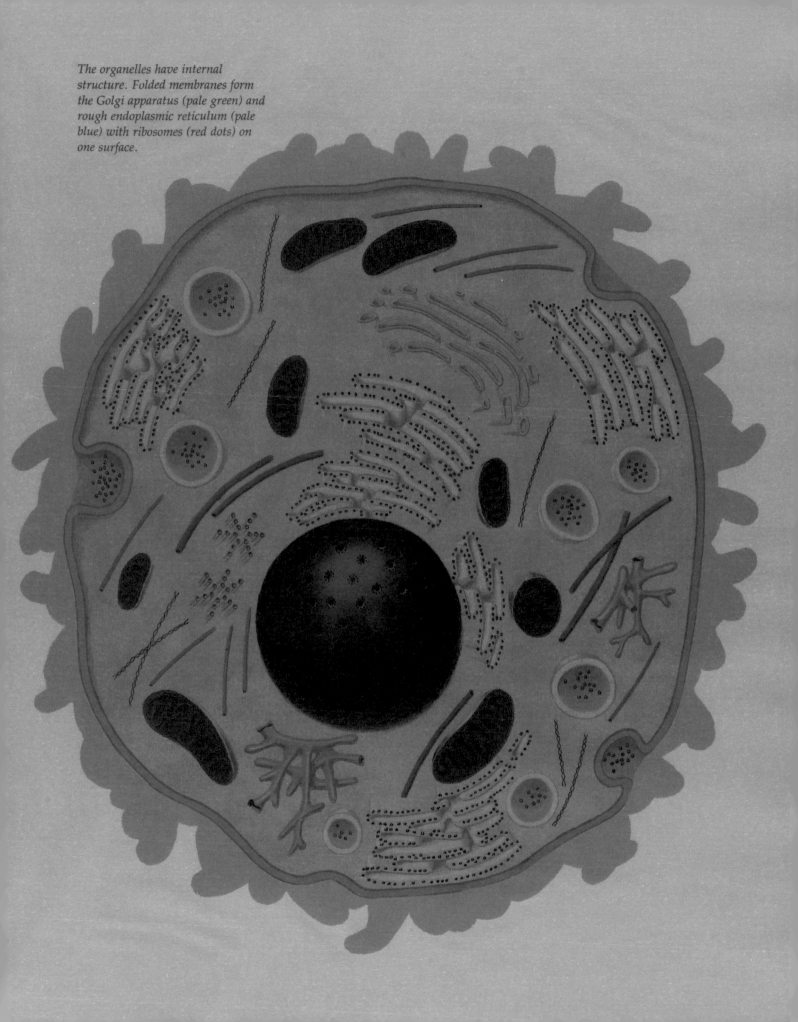

The organelles have internal structure. Folded membranes form the Golgi apparatus (pale green) and rough endoplasmic reticulum (pale blue) with ribosomes (red dots) on one surface.

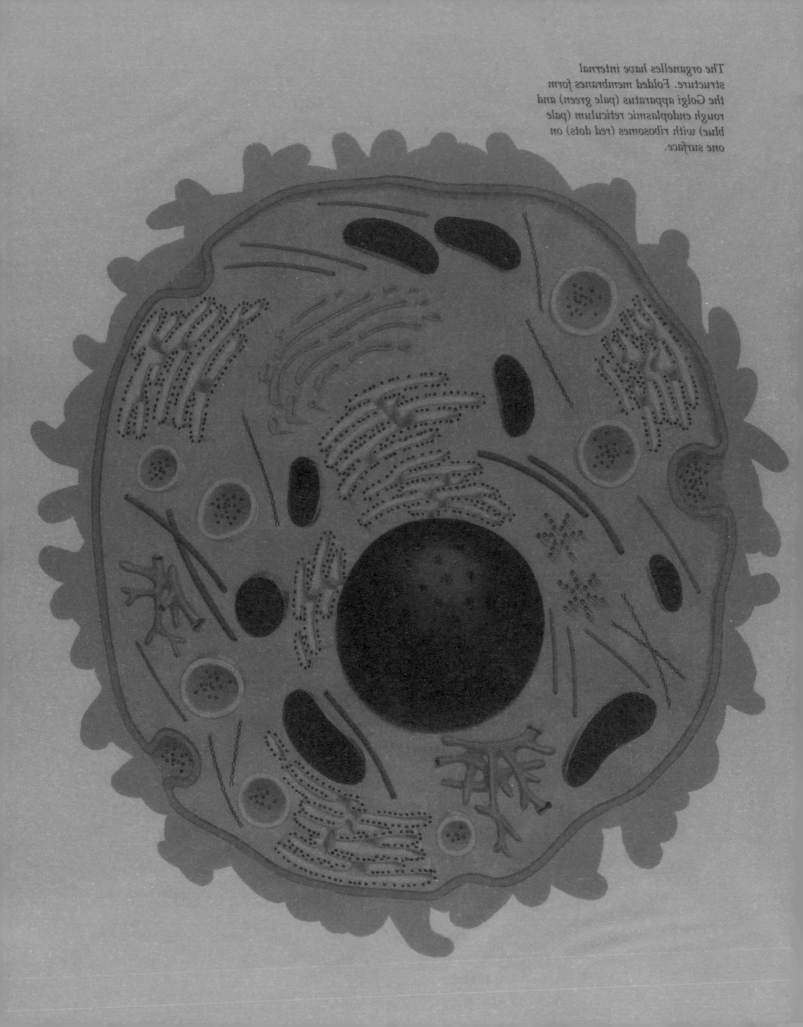

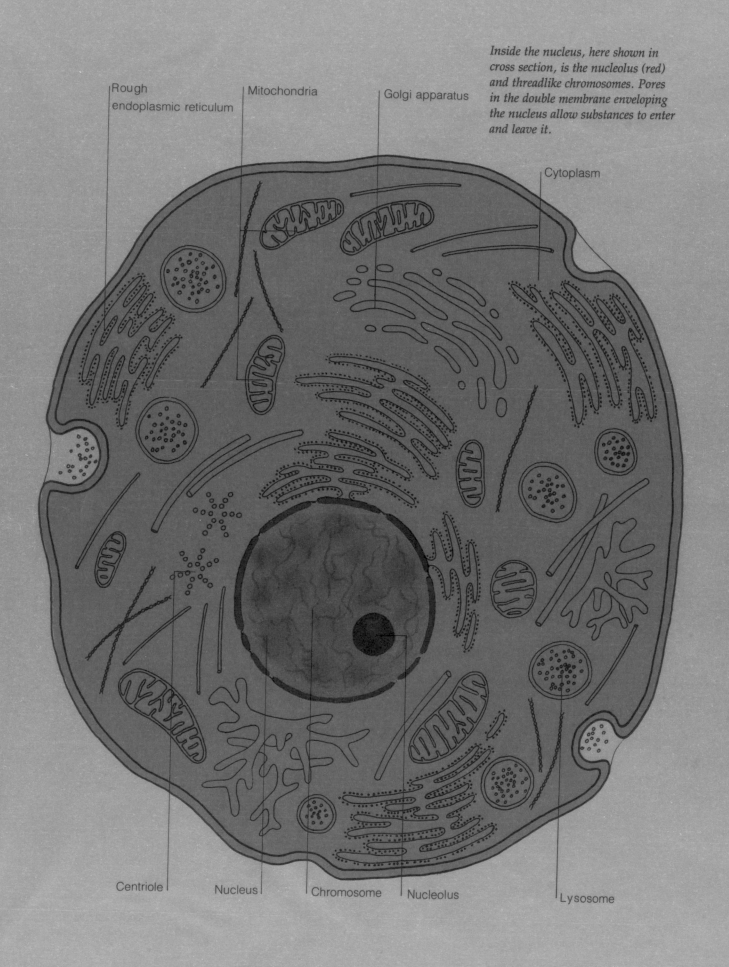

Rough
endoplasmic reticulum   Mitochondria   Golgi apparatus

*Inside the nucleus, here shown in
cross section, is the nucleolus (red)
and threadlike chromosomes. Pores
in the double membrane enveloping
the nucleus allow substances to enter
and leave it.*

Cytoplasm

Centriole   Nucleus   Chromosome   Nucleolus   Lysosome

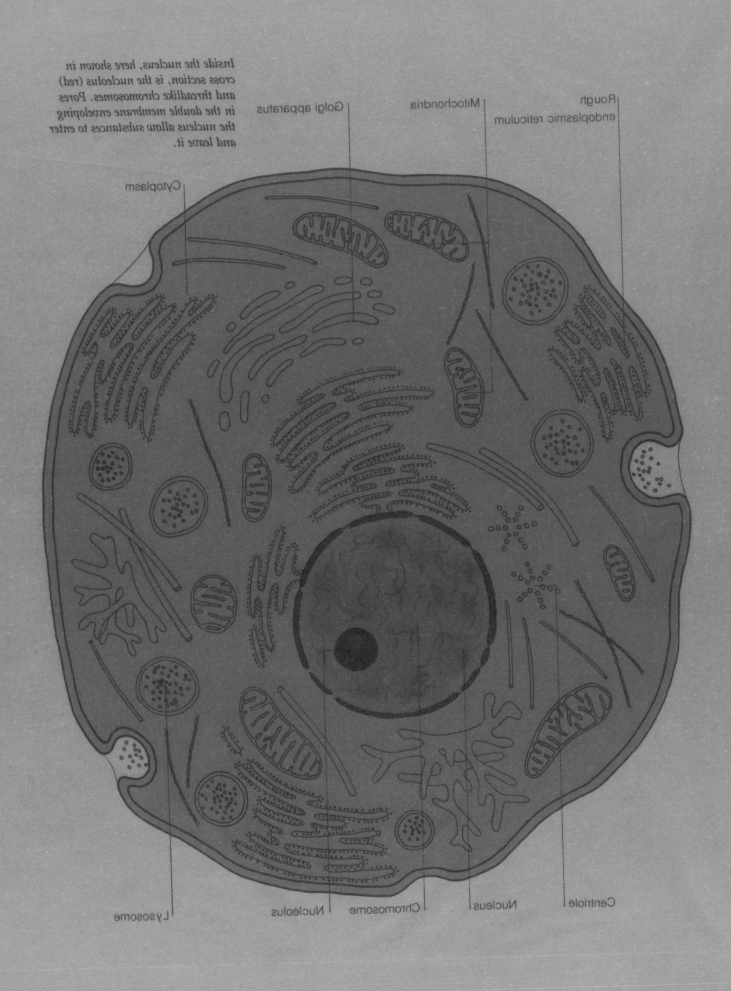

Chapter 3

The Specialized Cell

Cells have evolved in a myriad different ways to fulfil specific tasks within a complex organism such as a human being, or to exist within narrow ecological niches in the environment. In size, for example, the largest single animal cells were the yolks of the eggs of the extinct elephant bird, specimens of which have been found preserved in swamps; these fossil eggs have a capacity of about two gallons and are six times the size of ostrich eggs. In contrast, cells may be as tiny as those of the simple bacteria known as mycoplasmas, which are visible only under an electron microscope. Some mycoplasmas are less than eight millionths of an inch in diameter (many viruses are still smaller but are not true cells). And in the human body cell sizes vary between a motor neuron carrying messages from the spinal cord to the toes, which may be more than three feet long, to red blood cells, which are about three ten-thousandths of an inch in diameter.

An enormous degree of chemical variability exists within single cells. A simple mycoplasma may contain as few as seven to eight hundred proteins, whereas a complex protozoan animal may have ten thousand or more. Some simple bacteria living in hot springs have evolved proteins that remain active at temperatures high enough to destroy those of most cells. Other cells have adapted to environments that lack oxygen, and some can withstand great pressures, enabling them to survive at the bottom of the deepest oceans. This highly specialized metabolism could perhaps be harnessed to manufacture a wide range of chemicals.

In more complex organisms, individual cells are more limited in the tasks they perform. The cell becomes specialized in one type of function and loses other functions found in primitive cells. Single-celled organisms such as amebas must be able to carry out a wide range of functions as they move through their environment seeking food and

Ferdinand Lured by Ariel, *painted in 1849 by the British Pre-Raphaelite artist John Everett Millais, portrays a young man leaning forward to catch the ethereal whisperings of an air spirit. The sensory cells in the human ear can detect faint sounds but their range of sensitivity is limited to frequencies between about 20 and 20,000 hertz.*

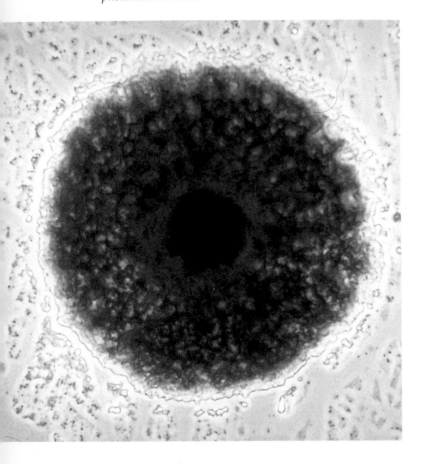

avoiding dangers. They have to be self-sufficient. But in complex organisms there are cells such as the red blood cells that have become so specialized they can carry out only one function: transporting respiratory gases (oxygen and carbon dioxide) around the body. In man, red blood cells are so highly adapted that they have few organelles and no nucleus. Other examples of specialization abound. Fingernails and hair, for instance, have a protective function, which they perform after their constituent cells have died — as do the superficial layers of the skin, which consist of the remnants of epidermal cells. Neurons, or nerve cells, have developed long intricate processes which connect with other neurons, and with muscle and other cells. The neurons form a complex communications network, transferring chemical and electrical signals to and from all parts of the body. Cells in the pancreas synthesize and secrete large quantities of the digestive enzymes required to break down food. Muscle cells have specialized structures (fibrils) which allow them to contract and numerous mitochondria — the cell's power plants — to drive them. In fact it is estimated that there are about two hundred different specialized types of cell in the human body.

Cell Classification

In organisms such as humans there are several ways of classifying different cell types into related families based on the cell's appearance and function in tissues. In one convenient classification four principal types of tissues are recognized: epithelial, connective, muscular and nervous. Epithelial tissue covers surfaces such as the outside of an animal and lines the digestive tract within. Included in this type are glandular cells, which secrete substances needed by the organism. Connective tissue includes fibrous tissue which holds the body together, the cells that make bone and cartilage, fat cells and blood cells. Muscular tissue consists of various types of muscle cells specialized to contract. Nervous tissue provides sensory input from the environment, integrates this input and directs and controls the reaction of the animal. Within each tissue type there are individual differences between closely related cells, enabling them to perform different functions.

The Nerve Cell

Neurons, or nerve cells, are among the most diverse cells in the body. They are designed to receive, conduct and transmit signals. To do this most neurons have three distinct portions: a cell body, an axon, and several dendrites. The cell body is the organizational center of the neuron and contains the nucleus, the Golgi apparatus, and the cell's protein-manufacturing center (the endoplasmic reticulum). Dendrites are projections that extend out of the cell body in all directions like the branches of a tree. Usually short, they receive signals from other neighboring neurons at contact points called synapses. The branching arrangement of the dendrites gives the cell a large surface area over which these signals may arrive. In addition, some dendrites extend a long way from the cell body and can pick up signals from more distant neurons.

The human body contains about 200 different types of cells. The diagram illustrates some of them and shows how they are arranged in different ways in the tissues of various major systems of the body.

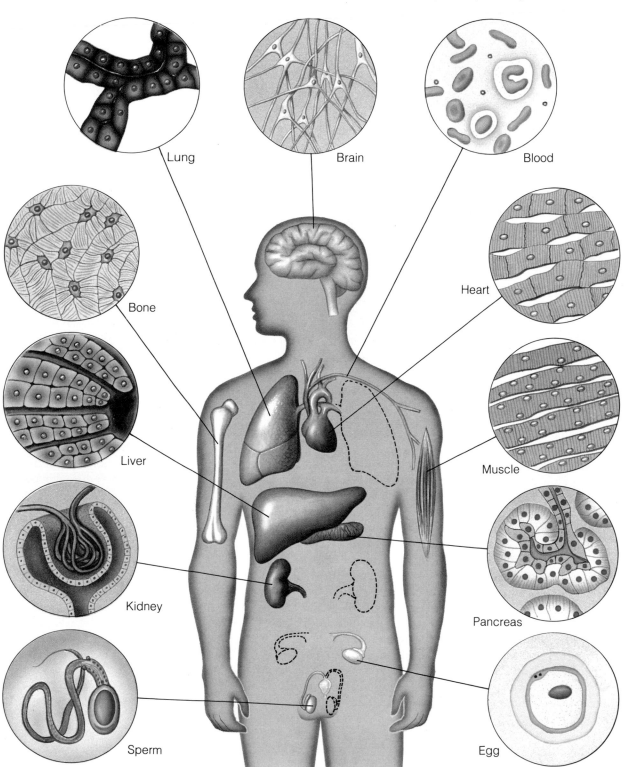

Lung

Brain

Blood

Bone

Heart

Liver

Muscle

Kidney

Pancreas

Sperm

Egg

Marie François Xavier Bichat

Pioneer of Histology

Bichat's recognition that the human body is made up of different types of tissues revolutionized the study of disease. His discoveries were gleaned from the many dissections he made. In one year alone he is recorded as having performed some 600 autopsies. Perhaps it was his constant contact with dead bodies that led him to remark rather pessimistically ''life is the ensemble of functions that resist death.'' He is remembered, however, not only for his pioneering work in pathology but also for founding histology, the study of tissues.

He was born at Thoirette in the Jura region of eastern France in November 1771. After studying humanities at Montpellier and philosophy at Lyons, Bichat changed to medicine, following in his father's footsteps. Under the patronage of Pierre Desault, a leading Parisian surgeon, he concentrated on surgery and anatomy, and established a reputation for his skill in treating fractures and in vascular surgery. He started a private course in anatomy and introduced the novelty of animal vivisections to teach his students physiology.

From his dissections, Bichat managed to distinguish between 21 tissue types on the basis of

differing textures and properties, such as extensibility and contractability. He realized that an organ may be composed of several different tissues, forming a ''web.''

He firmly believed that pathology should be based not on the topographical location of organs within the body, as had previously been thought, but on the composition of the organs themselves. In his preface to *Anatomie Générale* he stated that just as chemistry is the science of elementary bodies, anatomy is the science of elementary tissues, which differ from each other in composition and in the arrangement of fibers.

Interestingly, Bichat, like many people at that time, was skeptical of the microscope and did not use it to further his studies.

Interest in the organs of the body naturally led Bichat to investigate the effects of diseases on tissue. The commonly accepted belief at that time that a disease resided in the organs of the body was disclaimed by Bichat, who stated that disease only affected some of an organ's constituent tissues and that it was nothing more than some alteration to the tissue's vital properties. He believed that the great diversity of symptoms and diseases could be understood only by looking at the diversity of the tissues. Some properties of tissues, however, he believed to be irreducible to physical laws.

In 1800 Bichat was appointed physician at the Hôtel-Dieu in Lyons but two years later died from tuberculous meningitis, struck down at the age of 32 by the very subject of his research. Napoleon Bonaparte himself set up a bust in the Hôtel-Dien, mourning such a great contributor to eighteenth-century medicine. His pioneering work, however, had a lasting effect into the nineteenth century, not only in the fields of histology, cytology (the study of cells) and cellular pathology, but also in philosophy.

The axon is the nerve cell's transmitter. All of the information passed to the cell via the dendrites is processed into a signal which, if the input is strong enough, is passed along the axon until it reaches another cell. The axon has few organelles other than longitudinally arrayed microtubules known as neurotubules. Often an axon has numerous terminal branches so that when a signal arrives it can be passed to many different receiving cells. Thus a single neuron may ''focus'' the input from many different cells and distribute the message to numerous other cells, and this makes possible many and varied responses.

The outer membrane of a neuron is where signals are received and integrated. The membrane has a specialized sodium-potassium pump which actively pumps sodium ions out of the cell and potassium ions in, thereby separating the ions across the membrane. The ion separation produces an electrical charge across the membrane, called the membrane potential. Neuron membranes also contain specialized proteins which are very sensitive to the electrical charge. The proteins form into ''voltage-gated'' channels, which allow depolarization — the outward passage of potassium ions when the membrane potential of the cell changes. Rapid depolarization creates a local area of current flow known as the action potential, or nerve impulse. This current is the electrical basis of nerve signaling. Once one patch of membrane is depolarized, the electrical charge spreads along the cell surface, opening more voltage-gated channels, while those behind snap shut. In this way the nerve impulse travels along the axon until it reaches the terminal branches.

Many axons have an insulating sheath of myelin which speeds conduction of the nerve impulse. The myelin sheath is formed by a Schwann cell (or oligodendrocyte), which wraps itself round and round the axon like layers of insulating tape. When the wrappings of two Schwann cells meet, a gap is left where the axon membrane is left uncovered; this gap is the node of Ranvier. Because myelin acts as an insulator the nerve impulse cannot spread along the parts of the axon covered by the myelin sheath. Instead it ''jumps'' from one exposed part of the axon (that is, a node of Ranvier) to the next, where a new burst of current is produced. In this

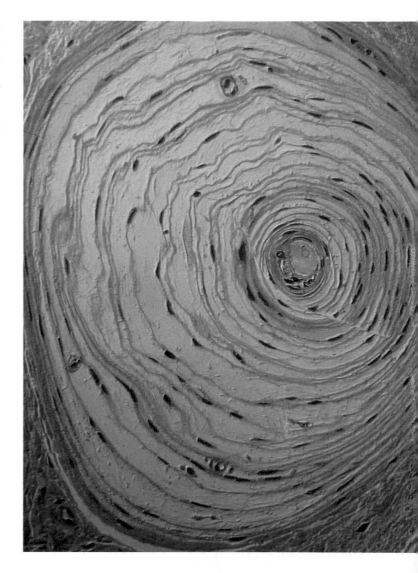

A Pacinian corpuscle (shown magnified about 250 times) is a sensory organ in the skin responsive to pressure, vibration and tension. It consists of roughly concentric layers of flattened cells and connective tissue (with a nerve fiber at the core), giving the characteristic onionlike appearance in cross section.

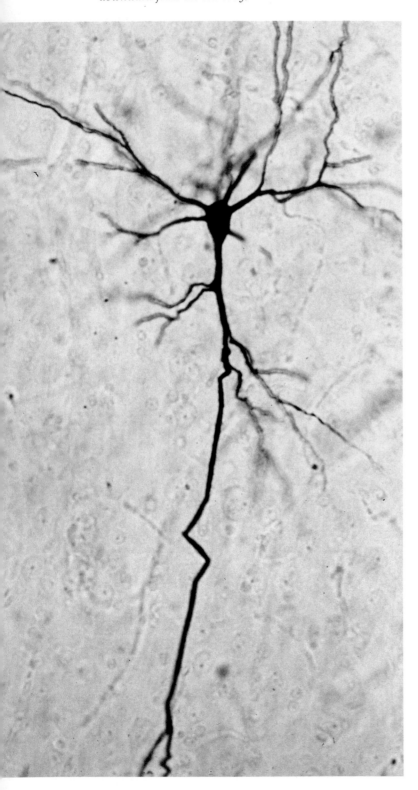

A typical nerve cell resembles a tree, with its branchlike arrangement of dendrites extending from the cell body (the oval structure toward the top) and a long axon (running downward from the cell body).

way the myelin sheath speeds the flow of nerve impulses. In myelinated axons nerve impulses can reach speeds of 300 feet per second (100 meters per second) or more, five times as fast as in the fastest unmyelinated axons.

Nerve impulses travel by a kind of knock-on effect, like the ripple of movement along a string of shunted boxcars. Impulses cross from one cell to the next at contact points called synapses, which consist of the end of the transmitting axon and the target region on the membrane of the cell receiving the message. The two are separated by a small gap known as the synaptic cleft. The impulse is transmitted across the synaptic cleft by special transmitter substances.

The terminal axon branch at a chemical synapse contains many vesicles, small sacs in the cell's cytoplasm that store one of several chemicals known as neurotransmitters. When a nerve impulse arrives at the branch, the vesicles fuse with the plasma membrane and release their contents into the synaptic cleft. The chemicals move across the synapse, meet up with a protein on the nerve cell on the other side, and stimulate it to generate another nerve impulse. The transmitter chemical is like a messenger boy who takes a written note from the sender to the receiver.

The Red Blood Cell

In sharp contrast to nerve cells, with their amazing complexity of cell processes and responses, the red blood cells (also called erythrocytes) have a strictly limited function. In normal circumstances mature human red blood cells are used solely to carry oxygen and carbon dioxide around the body to where they can be used or exhaled.

Red blood cells are formed in the bone marrow, and when immature they have all the organelles and biosynthetic machinery found in other cells. Under the influence of the hormone erythropoietin, immature red blood cells divide and begin to synthesize large quantities of the oxygen-carrying protein hemoglobin. Eventually the cell matures by stopping the production of RNA in the nucleus and then pushing the now unneeded nucleus out of the cell. The remaining organelles (including mitochondria and ribosomes) disintegrate, leaving only the outer membrane

56

The coordinated contraction and relaxation of the athletes' muscles (above) depend on nerve impulses from the central nervous system being transmitted to the appropriate muscles. This happens at motor end plates (right), where motor nerve cells connect (by means of chemical synapses) with muscle cells.

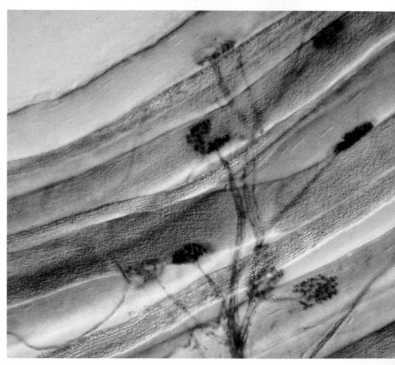

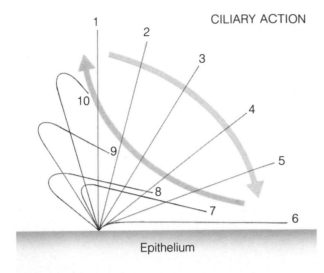

CILIARY ACTION

Epithelium

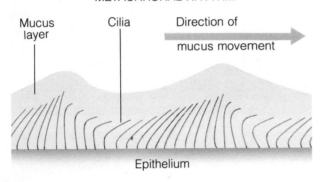

METACHRONAL RHYTHM

Mucus layer   Cilia   Direction of mucus movement

Epithelium

Cilia are hairlike structures used to move fluids or particles (or both). They achieve this by beating backward and forward. The upper diagram shows the movement of a single cilium, with the effective stroke in pink and the recovery stroke in purple. The lower diagram represents many cilia, showing the metachronal rhythm and the resultant movement of a fluid (mucus).

(with some associated proteins), hemoglobin, and a few enzymes for maintaining these two.

The mature red cell is then released into the bloodstream and pumped round the circulation (by the action of the heart), picking up oxygen in the lungs and releasing it to the tissues when it reaches the capillaries. In the circulation red cells gradually use up their store of enzymes and are damaged by being battered against each other and the walls of the blood vessels. As a result of these processes the red cells "age." They have an average lifespan of one hundred and twenty days, after which they are "eaten" by phagocytic cells in the spleen and elsewhere and destroyed.

Flagella and Cilia on Cells

Some cells have special hairlike structures that allow them to move through fluids or help to move fluids over their surface. These structures, although similar in many respects, have different names depending on their size and the number of them on a single cell. Flagella are relatively long and a cell usually has only one or two of them, whereas cilia are shorter and tend to be numerous.

Cilia move backward and forward in a coordinated pattern. A rhythm in which cilia are phased to beat slightly before or after their neighbors is said to be metachronal. Epithelial cells with cilia, such as those lining the Fallopian tubes, display a metachronal rhythm, one row of cilia moving slightly after those in front like the effect of a gust of wind moving across a wheat field. This produces regular waves of ciliary movement passing in the same direction over the cell surface. The wavelike motion helps to move substances and objects over the surface of the epithelium; for example, cilia move an egg cell from an ovary down the Fallopian tube to the womb.

Another well-known ciliary system is the one that operates in the respiratory tract. Here the movements of cilia sweep mucus, and bacteria and small particles trapped in it, up the air passages until they reach the windpipe. Coughing then removes the mucus from the respiratory tract. If this mechanism is disrupted or damaged, by smoking for example, then mucus and bacteria are not cleared from the lungs and may be a contributory factor in disease.

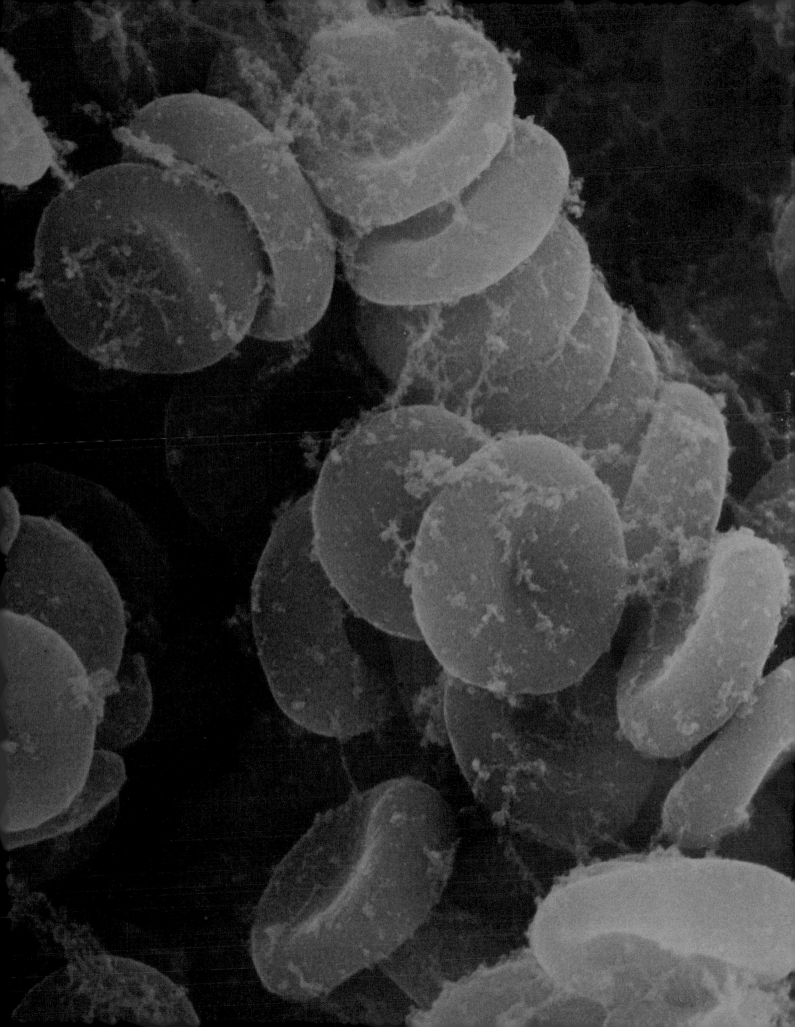

Sperm (bottom) use long, whiplike flagella to move about. Flagella have the same basic structure as cilia — the "nine-plus-two" arrangement of microtubules shown in the cross section (top).

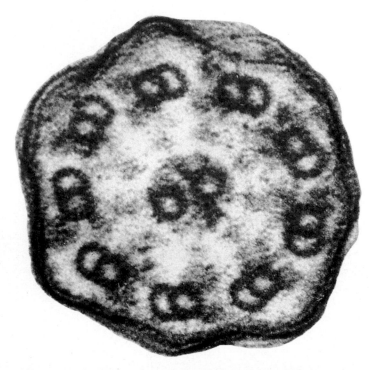

The long, whiplike flagella usually occur singly and are the means by which some single cells move themselves through fluids. The best known example in humans is the tail or flagellum of a spermatozoon. This moves in regular waves and propels the sperm along.

Both cilia and flagella contain a central structure, the axoneme, which provides the basis for their movement. The axoneme is a circular structure composed of two central and 18 peripheral microtubules (very fine tubes), the latter arranged in nine pairs. This "nine-plus-two" structure has been preserved throughout evolution, indicating that it provides an essential function. Electron micrographs seem to show that the outer microtubule doublets are attached to each other along their length by numerous regularly spaced arms, which gives the appearance of a rowboat with many oars. It is thought that these side (dynein) arms make the cilia move by sliding along adjacent doublets in a rowing motion. Because all the microtubules are fixed at their base (to a structure

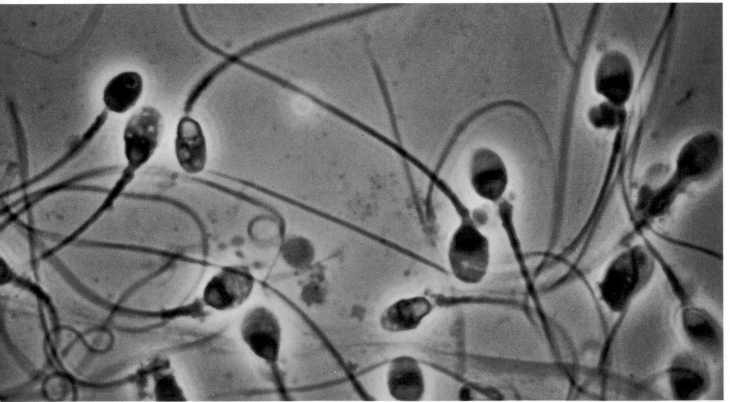

Specialized cells known as hair cells detect sound. Situated in the coiled cochlea of the inner ear, they rely on hairlike stereocilia (the long protrusions arranged in "fans" in the photomicrograph below) to

stimulate the auditory nerve cells to generate nerve impulses. The impulses travel to the cerebral cortex of the brain (the thin, convoluted outer layer of gray matter), where they are interpreted as sound.

called the basal body), when one side of the axoneme starts to "pull," the whole cilium bends.

There is a group of inherited diseases that may cause sterility in a man because the sperms are immobile. Study of defects in the sperm gives some insight into the importance of various parts of the axoneme. The immobility may be caused by defects in the dynein arms or in the central tubules, or the proteins surrounding the central tubules may be abnormal. As well as nonmotile sperm, those affected have defective cilia in their respiratory tracts and often suffer from chronic respiratory tract infections. Most interestingly, about half the individuals with abnormal cilia syndromes also have their internal organs reversed left to right, so that the heart is on the right side of the chest. This leads scientists to suggest that perhaps the structures that control cilia and sperm flagella also control movement of cells in early embryos. Abnormalities of movement might lead the primitive organs to end up on the wrong side of the body.

Cells that detect light, odor and sound all do so using modified cilia, or cilialike structures, as part of their sensory apparatus.

The Cells of Hearing

Sound is detected by hair cells in the cochlea, the snailshell-like part of the inner ear. They sit on a shelf of tissue called the basilar membrane, which separates them from a fluid-filled cavity. The upper ends of the cells are equipped with hairlike projections or stereocilia which, unlike true cilia, are filled with actin fibers (rather than the microtubules described earlier). Their tips are embedded in the tectorial membrane above them. As a result, any movement in the basilar membrane below the cell flexes the stereocilia.

Sound waves conducted as pressure ripples through the fluid in the cochlea thus cause the basilar membrane to move, so that the vibration of sound is converted into movement of the stereocilia. This movement in turn causes the hair cells to release neurotransmitter chemicals (from vesicles at the cells' bases) which act on nerve endings alongside the hair cell. Receptors on the nerve endings take up the neurotransmitter, electrifying the plasma membrane and generating a nerve impulse, which is transmitted along the nerve fiber to the brain and interpreted there as the sensation we recognize as sound.

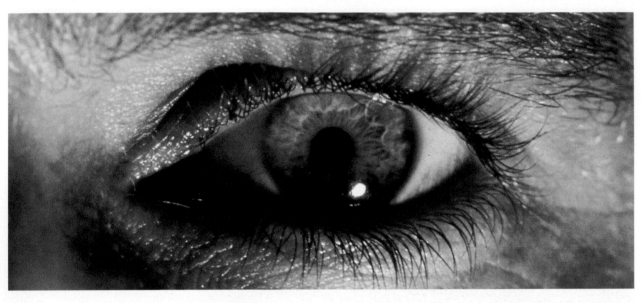

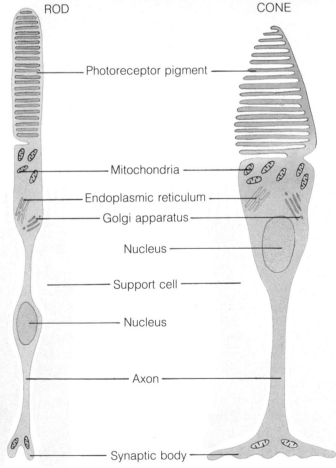

ROD   CONE

Photoreceptor pigment

Mitochondria

Endoplasmic reticulum

Golgi apparatus

Nucleus

Support cell

Nucleus

Axon

Synaptic body

The Cells of the Eye

Rod and cone cells in the retina at the back of the eyeball detect light. Cone cells detect colors whereas rod cells detect only light and dark but can do so in very dim light. Rod cells are so sensitive that they can respond to a single photon (a "particle" of light), although five or more photons must hit the human retina before anything is actually seen. In electron micrographs rod cells look like fingers with distinct segments and joints. The outer segment, the tip of the finger, contains the photoreceptor membrane, which looks like a stack of disks or pancakes. A modified cilium connects it to the inner segment, the middle of the finger, which contains mitochondria and Golgi apparatus. The rest of the cell, the base of the finger, consists of the part of the inner segment containing the nucleus, and is joined to a synaptic body at the base of the cell.

The membrane that contains photoreceptor proteins is synthesized like ordinary membrane in the endoplasmic reticulum of the inner segment of the rod. It is then passed to the bottom of the outer segment where it joins the membrane stack. When a photon of light hits the densely packed membranes there is a high chance that it will hit a photoreceptor protein (rhodopsin) associated with the membranes. The light energy makes the rhodopsin molecule change shape, causing the sodium channels in the cell's plasma membrane to close. This action, in turn, leads to the generation of a nerve impulse. Trains of impulses are processed by many neurons as they travel along the optic nerve to the brain, where they give rise to the sensation of sight.

The Cells of the Nose

Olfactory (smell) cells also use modified cilia, in this case to trap odor molecules. The cilia have the typical nine-plus-two tubular array at their center but appear to have odor receptor molecules in their plasma membrane. They hang in the nasal space like the tendrils of a sea anemone and sample the air flowing over them for odor-causing molecules. Olfactory cells are unusual in that they are one of the few parts of the nervous system that have the ability to replace themselves. New olfactory cells are made continuously in the nasal mucosa to replace degenerate or damaged neurons. And unlike visual and auditory receptor cells, each olfactory cell is directly connected to the brain by a long axon which passes through the skull and the roof of the nose.

The Cell Surface

All interactions with the environment and other cells occur at the cell surface — that is, via the plasma membrane and its attached components. Descriptions of generalized cells often give the impression that they have a smooth surface resembling a billiard ball or balloon. In fact this is not true for most cells, as is apparent when they are examined using the specialized technique of scanning electron microscopy. Viewed this way, most cells resemble a porcupine rather than a billiard ball, and some have truly fantastic surface specializations.

The cell membrane is rarely smooth. It may be thrown into folds or ridges, or numerous small fingerlike projections (microvilli) may develop. The outer surface of the plasma membrane is often the site of a "fuzz" of glycoprotein molecules projecting from the cell membrane.

Most cells use all of these methods to a greater or lesser extent. The range of variation is enormous. Red blood cells, which probably have the least complex surface, have a smooth biconcave disk

Deep inside the nose are the olfactory cells that sense smells. These cells use modified cilia — which hang from the roof of the nasal cavity — to trap odor molecules, thereby enabling the sensory cells to send "smell impulses" directly to the brain.

Microvilli are short, fingerlike processes that project from many cell membranes, such as those lining the kidney tubules (below). Their primary function is to increase the effective surface area of the cells.

The Englishman Daniel Lambert (1770–1809) continually put on weight throughout his short life, weighing 739 pounds at his death. But, remarkably, the number of fat cells in his body remained constant.

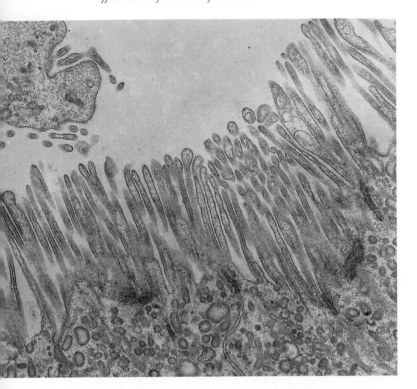

shape but possess a glycoprotein fuzz which includes the sugar molecules that determine an individual's blood group. Some brain cells have a treelike pattern of surface projections or dendrites, which increase the surface area of the cells enormously and allow them to make thousands of contacts with other cells.

The short fingerlike projections called microvilli are a common feature of cell membranes. Microvilli have an actin fiber core and can function in a number of ways. Some cells, such as white blood cells, send them out at random and it is thought that they may be used like "feelers" to sample the environment for the presence of potentially dangerous material such as bacteria. On other cells, for example the absorptive cells that line the small intestine, or the tubules of the kidney, hundreds of tightly-packed microvilli are always present. By being there they increase the available surface area of the cell by as much as forty times. In the gastrointestinal tract many digestive enzymes are embedded in the membranes of the microvilli so that the enlarged surface enhances the breakdown of large molecules as well as increasing the total amount of absorption of their digestion products. Gut microvilli are also coated with a special type of antibody molecule, IgA, which helps to block invasion of the body by disease-causing bacteria and viruses. By increasing the area coated with IgA, the chances of trapping an invader are greatly increased.

Energy Storage Cells

An organism that cannot store energy must spend nearly all of its time eating. But if an organism is to develop more varied behavior, it must find a way of storing energy to leave time for other specialized activities. Two types of storage systems have evolved to deal with this need, and both are found in humans. Certain cells store glycogen, a polymer of sugar molecules also known as animal starch, and highly specialized cells store fat.

Glycogen occurs in the largest quantities in liver and muscle cells, where it can be rapidly broken down to sugar (glucose). From the liver the sugar is released into the bloodstream to provide energy throughout the body. In muscle cells the sugar is usually consumed within the cells themselves. But glycogen is a relatively inefficient energy store and fats produce proportionately about six times more energy when burned as fuel than does glycogen. If glycogen were the only energy store, most people would have to weigh about 50 pounds more than they do by storing energy as fat.

Fat is stored in special cells called adipocytes, which are among the largest in the body. A fat cell has a nucleus which is squashed to one side and a thin rim of cytoplasm containing all the usual organelles. The rest of the space is taken up by a large fat droplet, which accounts for about 95 percent of the cell volume.

A fat or lipid droplet is formed by fusion of many smaller droplets synthesized in the narrow but extremely active rim of cytoplasm beneath the cell membrane. An adipocyte has receptors which bind the specific lipoproteins that transport fats in the bloodstream. Receptor-bound lipid is taken up by endocytosis in vesicles and then broken down in the cytoplasm to produce the building blocks needed to make a storage droplet. Because fats are not soluble in water there is no need for the droplet to be separated from the cytoplasm by a membrane;

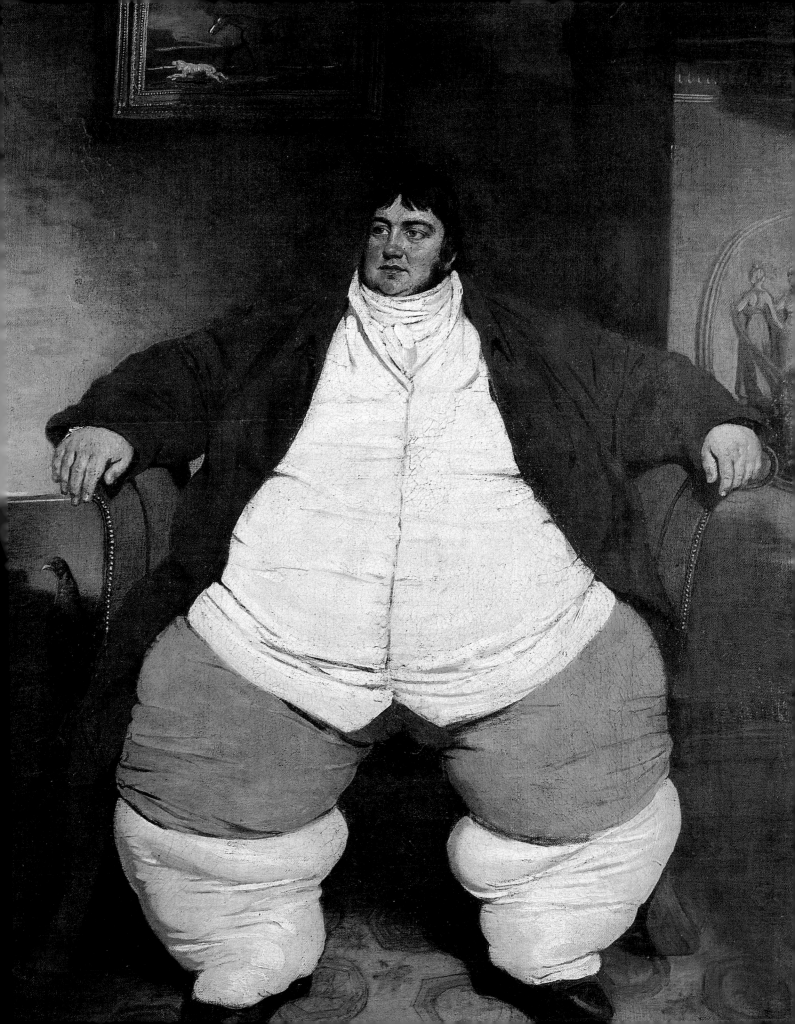

This portrait of the Virgin and Child epitomizes one of the female body's adaptations to motherhood, namely the production of milk from groups of specialized secretory cells within the breasts.

The main types of exocrine gland are shown below, with their secretory parts in red. Simple saccular, or sebaceous, glands secrete sebum, which lubricates the hair and skin. Simple tubular glands are found in the walls of the stomach. Sweat glands are an example of simple coiled glands. Salivary glands are an example of compound saccular glands, while compound tubular glands occur in the duodenum.

it merely pushes the cytoplasm aside until the cell resembles a fluid-filled balloon.

The Secreting Cell

Many cells have become specialized to provide secretions that serve many different functions. Tears, for example, help to lubricate the eye and protect it from infection; digestive enzymes break down food in the gut; and many types of hormones are released into the bloodstream and act as chemical messengers, transmitting information round the body and stimulating appropriate responses from their target cells.

Two major kinds of secretory activity exist — endocrine and exocrine. Most endocrine secretions are hormones and are released from the producing cell directly into the bloodstream, and thence carried to their destination. Exocrine cells release their secretions into tubes or ducts, which carry the secretions to a surface. One of the simplest types of exocrine structure is a sweat gland, which produces large amounts of salty fluid when stimulated. Sweat is carried to the skin surface by sweat ducts. Many other fluid secretions are produced in an essentially similar way.

The production of milk by the breast requires one or two modifications, however, because milk contains protein and fat as well as water and salts, although it may be that milk glands are in fact derived from sweat glands. Milk proteins are produced on rough endoplasmic reticulum (as are all secreted proteins), and are then packaged into vesicles in the Golgi apparatus. The contents of these vesicles are released from the cell by exocytosis. Milk fat is synthesized within the cell as droplets, which move through the cytoplasm to reach the cell surface. The droplet appears to push the membrane out to form projections which then break off as membrane-covered drops. In other words, the milk-secreting cell actually releases part of itself around the fat droplet.

Digestive enzymes are proteins made in exocrine (acinar) cells of the pancreas, and also in many other cells. The pancreatic cells have the most highly developed rough endoplasmic reticulum of any in the body. Enzymes are synthesized in the same way as any other secretory protein and packed into "packets" (vesicles) called zymogen

PRINCIPAL TYPES OF EXOCRINE GLANDS

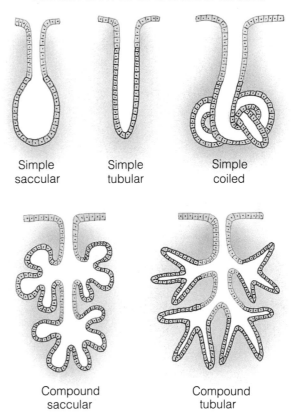

Simple saccular

Simple tubular

Simple coiled

Compound saccular

Compound tubular

granules. Because active digestive enzymes inside the cell would be harmful to many metabolic processes, they are synthesized as inactive molecules or zymogens and stored in this safe form. The granules pack the cytoplasm of the pancreatic acinar cells next to the duct that will carry secretions to the small intestine. When a chemical message indicating the presence of food in the duodenum is received by the acinar cell, its granules fuse with the surface, releasing their contents of digestive enzymes. The inactive enzymes are activated by breaking off small portions of their length, but only after they have been released from the cell that made them.

Endocrine secretions are released directly into the bloodstream. Cells that make protein hormones, such as insulin or those secreted by the pituitary gland, do so using the mechanism of protein synthesis, packaging the hormones into intracellular vesicles for storage and later releasing

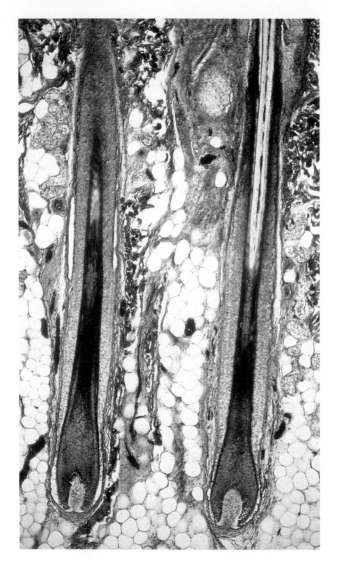

The cross section through the skin of the scalp shows two hair follicles surrounded by fat cells (adipocytes) in the layer of subcutaneous fat below the dermis (the main layer beneath the outer epidermis).

them by exocytosis — as in exocrine secretory cells. Steroid hormone producing cells — the Leydig cells that produce testosterone in the testes, for example — have a different specialized structure. To synthesize steroids they use smooth rather than rough endoplasmic reticulum, because steroids are produced by the actions of specific enzymes on fats (lipids, especially cholesterol) rather than by protein synthesis. Many of the enzymes required are associated with the membrane of the smooth endoplasmic reticulum. Unlike many protein hormones, steroids are usually released as soon as they are made, and for this reason there are no storage vesicles in steroid-forming cells.

How do Cells Form Organs?

Although details of how a particular specialized cell performs its functions are often well understood, there remains a fundamental mystery surrounding the subject of all specialization. How do cells become specialized and form themselves into correctly positioned organs with well-defined functions?

All the cells in a complex organism — whether they are in the liver, brain or teeth — are the descendants of a single fertilized egg. What processes regulate the transition from a cell with many possible descendants to a cell whose future development is fixed? How do cells that start as part of the unspecialized group of cells in an early embryo know where to go as the embryo develops? This is the study of differentiation — how cells become different and specialized — and it remains one of the most poorly understood topics in cell biology.

There are a few clues that suggest how some aspects of differentiation occur. Germ cells — the forerunners of egg and sperm cells — migrate through the embryo. For reasons that are not clear those that go to the wrong places die, and it appears that the connective tissue of the "right" place in the developing gonad specifically nourishes the primitive germ cell.

Motor nerve cells seek out and attach to muscle cells in the developing embryo, and in some cases the "feeler" (the axon or dendrite) sent out by the nerve cell body must grow several inches before reaching its destination. In this case it appears that the muscle cell sends out a chemical that attracts the growing tip of the nerve like a scent for a bloodhound to follow. If a growing nerve tip fails to find a muscle cell and form a synapse, the whole nerve cell dies.

The process by which a chemical substance attracts a particular cell to a certain place is known as chemotaxis. One example of this mysterious process occurs when the body is injured or is subject to local inflammation: leukocytes (white blood cells) flock immediately to the site as a response, as the body takes primary steps to heal itself as rapidly as possible. Another example — which occurs in plants, though not in the higher animals — is constituted by their eggs, which secrete a certain acid in order to attract sperms toward themselves. The substances that are the principal agents of chemotaxis (which are small molecules) bind to receptors on the blood cell membrane and direct it toward the stimulant.

Bodybuilders aim to increase the proportion of body muscle to fat so that their muscles are well defined. This aim can be achieved by a combination of exercise, which makes the muscles larger, and proper diet.

Cells with similar properties may specifically adhere to each other, which suggests that they have molecules on their surface that make them "sticky" for each other but not for other cells. The carbohydrate component of many membrane glycoproteins is a strong candidate for this role. Thus a cell that makes contact with a cell of similar stickiness remains in that site, and gradually a clump of similar cells is built up. Furthermore, when such cells divide they tend to stay in one place rather than migrate away.

In general terms, cells contain their own special internal program determined by their genetic material, and this program is initially the same for every cell. As a result of program changes, yet to be explained, the cell is directed to produce specific membrane proteins called receptors, which can bind to small molecules released by other cells. When a receptor binds to the correct molecule it can give the cell a chemical message which causes a previously quiet part of the program to become active. If it stimulates the cell to make a different receptor, the cell is then able to detect the presence of other signal molecules which may make the cell divide. This type of chemical signaling between cells may be one of the factors responsible for differentiation.

Some differentiated cells have lost the ability to divide. Once they are destroyed they are lost to the body forever. One goal of the study of differentiation is to understand how specific cells are switched on and off so that lost cells could be replaced by cells from the same organism. What is known as the chalone concept attributes to each type of tissue a so far untraced and unidentified substance which has as its specific function the "maturing" of cells, with the result that they then die.

This idea of having a "bank" of organ tissues that would not be rejected by the body because they were identical to the patient's own cells may seem like science fiction. But the method is already being used to treat burns. Cells from an unaffected part of the skin of the patient are cultured as a tissue under laboratory conditions, then laid across the burn, where they continue to multiply until the damaged area is covered with new skin. The cells behave exactly like a skin graft taken from another part of the patient's own body.

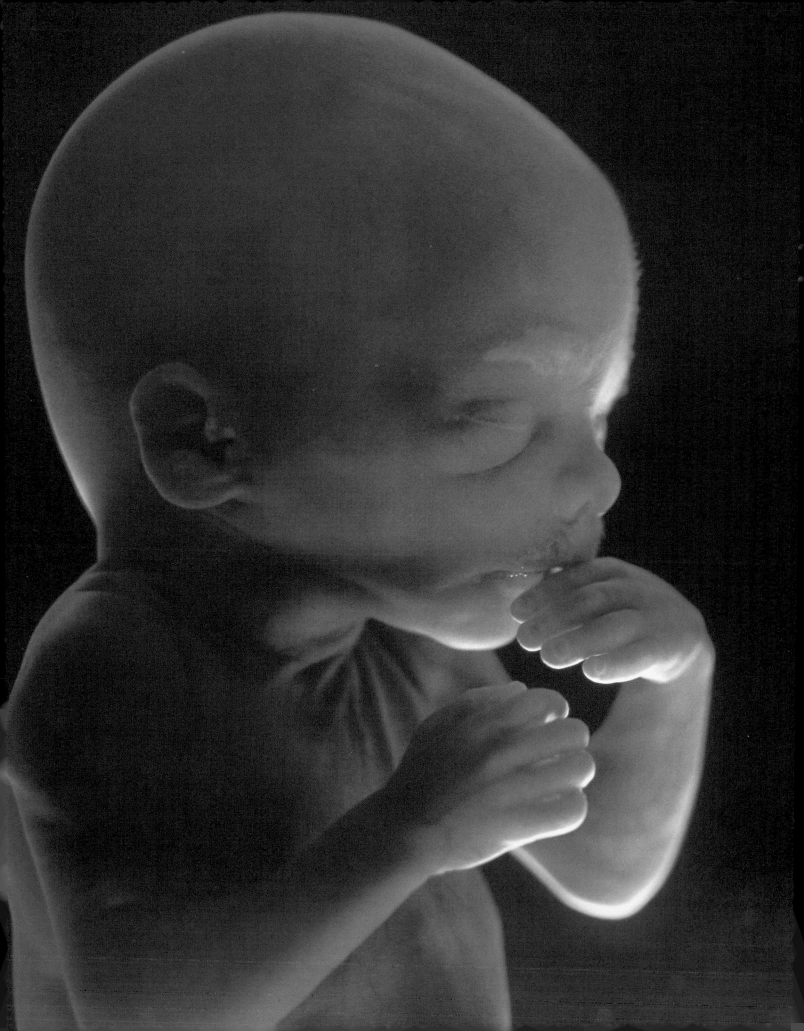

Chapter 4

How Life Grows

No natural force is more indicative of life's wondrous power than growth. In precisely timed and orchestrated sequences, the cells that make up all living organisms divide in two, with each daughter cell capable of transmitting the genetic messages of its progenitor.

The body of an adult human contains more than 200 different types of cells, and these can be categorized into three main groups according to their basic pattern of replication. Some cells, such as striated muscle cells and neurons, usually do not reproduce, although they have a limited capacity to repair damage to individual cells. For example, some nerves can regenerate after injury if their cell bodies (the part that contains the nucleus, the substructure in which the cell's genetic material is situated) remain intact.

The second group of cells, such as hepatocytes (which make up the bulk of the liver), have a moderate capacity to divide if cellular tissue is lost through injury, surgery or disease.

The third group of cells, for example those of the hematopoietic system (which produces blood cells) and the skin, divide actively to compensate for a relatively short life or high rate of natural loss. The skin cells are continually being shed from the surface of the skin as part of the normal protective process and these have to be replaced at the basal layer by regular cell division. Similarly, red blood cells, which lose their nuclei prior to being discharged into the bloodstream, have a lifespan of about 120 days, after which time they are removed from circulation by macrophages (large cells that have the ability to engulf other cells) in the liver and spleen. In order to keep up with this process a regular supply of red blood cells is being made.

An insight into how cells reproduce and tissues grow was not achieved until the mid-nineteenth century, when microscopic techniques enabled scientists to examine the cells closely. It became clear that for satisfactory growth of a tissue (or

The miracle of growth is nowhere more astonishing than in the creation of a new human being. Over a timespan of only 40 weeks (the average length of pregnancy) a single cell — the fertilized egg — grows and replicates many times, developing into an embryo, then a fetus, and finally into a perfect miniature person.

71

Hydra consist of only a few different types of cells and can reproduce merely by budding off new individuals. Human beings have about 200 types of cells and require a longer, more complex development.

The fine threads of the DNA molecule are normally coiled in a tangled mass inside the cell but have been spewed out of the sausage-shaped bacteria and so are visible in the electron micrograph on the right.

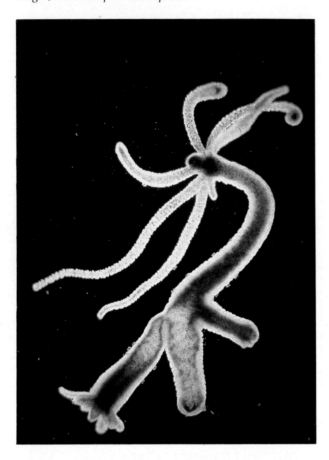

which along with protein makes up the bulk of the cell nucleus, contains the genetic message that enables each cell to produce a replica of itself, and also provides the information the cell needs to carry out its other functions.

In the late 1940s researchers began to speculate about the odd and characteristic shape they had seen in X-ray studies, calling DNA "The scroll upon which is written the pattern of life." In 1953 an American investigator named James Watson and a British researcher, Francis Crick — working together at Cambridge University, England — announced that they had ascertained the structure of the elusive DNA molecule. Aided by the X-ray diffraction work of Maurice Wilkins and Rosalind Franklin, Watson and Crick discovered that DNA was in the shape of a double helix. In describing their revolutionary model Watson and Crick stated, "We should like to propose that each of our complementary DNA chains serves as a template for the formation onto itself of a new companion chain." Thus their model solved an important problem. It provided a possible explanation of how DNA might replicate itself and endow the cells of the next generation with the exact genetic components of the previous one.

Subsequent work has confirmed the theory of Watson and Crick and a workable understanding of the biochemistry of DNA is now available. The double helix is like a twisted rope ladder. Each of the two side ropes is a single strand made up of alternate sugar molecules (deoxyribose) and phosphate molecules. If one imagines that the sugar molecules are the knots between the rungs and the side rope, then the phosphate molecules are the spare rope between the knots. To each of the sugar molecules is attached a nitrogen base molecule (imagine a half-rung on the ladder), of which there are four: adenine and guanine (which are purine nitrogen bases); thymine and cytosine (which are pyrimidine nitrogen bases). Each single unit made up of the nitrogen base (half a rung), a sugar (the knot) and a phosphate (the spare rope) is called a nucleotide.

The half rungs of each strand of the rope ladder are linked together in the middle with chemical bonds called hydrogen bonds. But these links are not random: adenine always links (by means of two

repair of cellular damage) the cell needs to reproduce itself. In this way each specialized cell line is perpetuated and can continue to perform the function of its ancestors.

This is a general rule, but there are certain cells, such as white blood cells and primitive stem cells of the hematopoietic system, that can specialize in various ways depending on the influence of external factors. For example, a certain type of white blood cell, the monocyte, can undergo transformation into a scavenger cell known as a macrophage which devours foreign material. But this happens only if the monocyte is stimulated by local infection or damage.

The DNA Molecule

One of the most important biological discoveries of the twentieth century has been the identification and understanding of the structure and function of DNA (deoxyribonucleic acid). This compound,

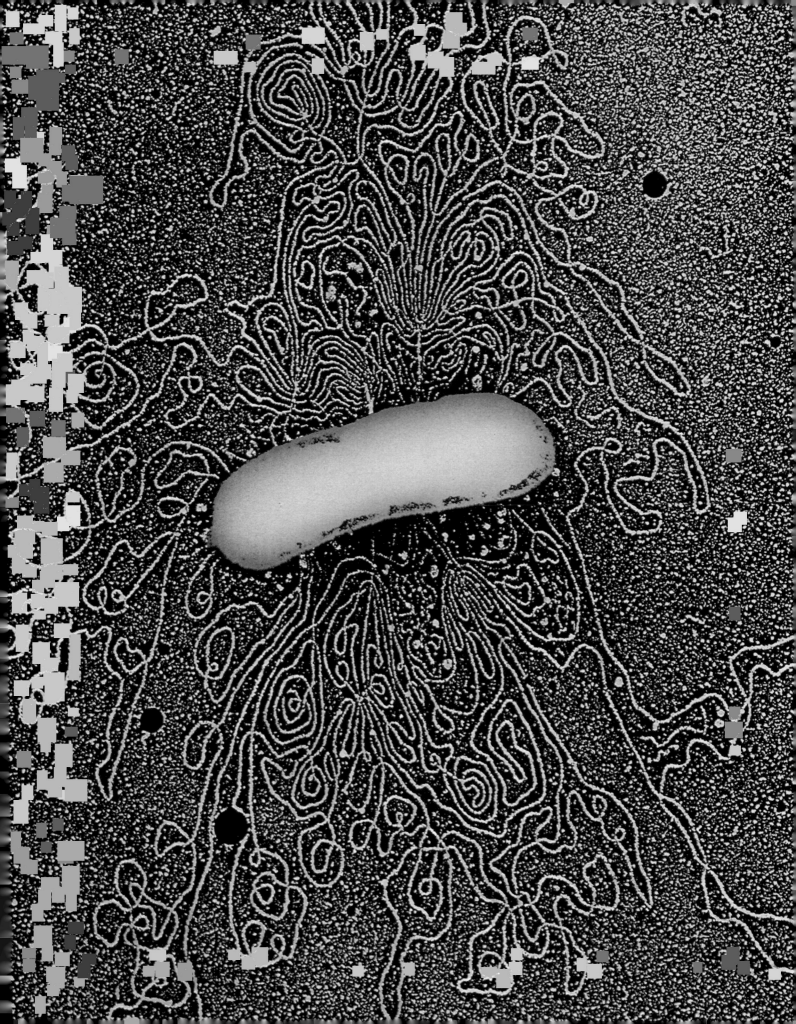

The DNA molecule—the cell's genetic material—has a double helix shape. It consists of two twisted side chains of sugar (pale green) and phosphate (purple) molecules, with cross-links of nitrogen base molecules; these are adenine (dark green), thymine (orange), cytosine (red) and guanine (blue). The bases can link together in only one way: adenine with thymine, and cytosine with guanine. The upper part of the diagram represents this arrangement schematically; the lower part depicts the shapes of the molecules more accurately and also shows the hydrogen bonds (black dashes) that constitute the links between bases.

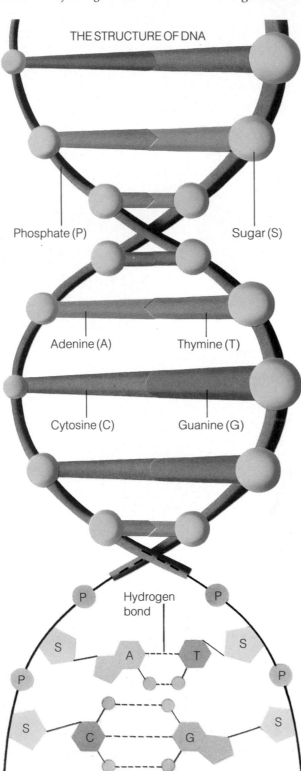

THE STRUCTURE OF DNA

Phosphate (P)

Sugar (S)

Adenine (A)

Thymine (T)

Cytosine (C)

Guanine (G)

Hydrogen bond

P   P

S   S

A   T

P   P

S   S

C   G

hydrogen bonds) to thymine and guanine always links (with three bonds) to cytosine, so one strand of the double helix is an exact complementary copy of the other.

DNA is an extremely large molecule and each strand has many thousands of nitrogen bases. It is the order of these bases along the strands that is the basis of the genetic code by which protein synthesis in the cell is controlled (a process described more fully in Chapter 2). And so, because the types of protein synthesized determines the type and activities of the cell, DNA is the ultimate store of the information necessary for all cell functions.

DNA Replication and Cell Division

As a preliminary to cell division the DNA strand of genetic material has to replicate exactly. It reproduces itself for the nuclei of the two daughter cells. The hydrogen bonds between the nitrogen bases break down, the rungs halve, and free nucleotides attach to their free ends. Because thymine bonds only to adenine and guanine bonds only to cytosine, two identical strands result.

Once replication has occurred, the cell contains twice the normal amount of genetic material (i.e. twice the number of DNA molecules). The nuclear membrane then breaks down and the DNA condenses into distinct strands visible under the light microscope by using certain forms of staining or special techniques such as phase contrast. These strands, wrapped in protein, are the chromosomes; each chromosome contains two strands of DNA. The chromosomes then split in two and the resultant single strands of DNA move to opposite ends of the cell, which subsequently divides in two, each with its own complement of DNA. In this way two new identical cells are formed, each with the normal amount of genetic material. The strands then unravel in the new cell nucleus.

The whole process of cell replication—known as mitosis—has been the subject of considerable research and is understood in some depth. It results in cells with the full complement of genetic material. A specialized type of division, meiosis, produces eggs and sperm—which have only half the full complement of genetic material, the egg's from the female (mother) and the sperm's from the male (father).

Before a body cell can divide, its DNA must replicate so that each of the future two daughter cells will have the full complement of genetic material. The replication process is illustrated in the diagram below.

First, the links between the nitrogen bases break — in effect the "parent" DNA double helix unzips. Then free nucleotides (each consisting of a sugar molecule, a phosphate molecule and a nitrogen base) link

up with the exposed ends of the bases on the parent DNA. This process continues until two complete "daughter" DNA molecules have been formed, both exact duplicates of the parent DNA molecule.

Mitosis

The process of mitosis can be subdivided into several stages, called prophase, metaphase, anaphase and telophase. In the period between active cellular mitosis, called interphase, the nuclear material is confined within its nuclear membrane as an amorphous mass. It is during this phase that DNA replication occurs, and this results in two identical DNA molecules. Once DNA replication is complete the visible mitotic phase is ready to start (although it does not usually do so immediately), beginning with prophase and ending with telophase.

Prophase to Telophase

In prophase the nuclear membrane dissolves and the nuclear material called chromatin becomes distinct as long thin threads (*mitos* means thread in Greek, thus mitosis). These structures — the chromosomes — then condense and thicken. The number of chromosomes is specific to individual species of animals and plants. In humans the number is 46, of which 44 are autosomes common to men and women and two are sex chromosomes which, along with other functions, determine the sex of an individual. (Each chromosome can be recognized by its shape and staining characteristics.) At prophase, because the genetic material has replicated during interphase, the cell contains twice the normal number of chromosomes (46 pairs in men and women).

During metaphase the 46 pairs of chromosomes line up along the equator of the cell. At this stage two small organelles — the centrioles —are visible at either end of the cell. From each centriole extends a series of fine fibers to each chromosome. These are the spindle fibers, which are made of microtubules (very fine tubes). They are responsible for the subsequent separation and movement of the chromosome pairs.

Every chromosome consists of a pair of identical chromatids. The chromatids are joined at some point along their length; this point is called the centromere. Each chromosome is a different length from the others and the point of attachment varies, making them easy to identify as individuals. And because the individual chromosomes can be identified at this stage of mitosis, doctors and

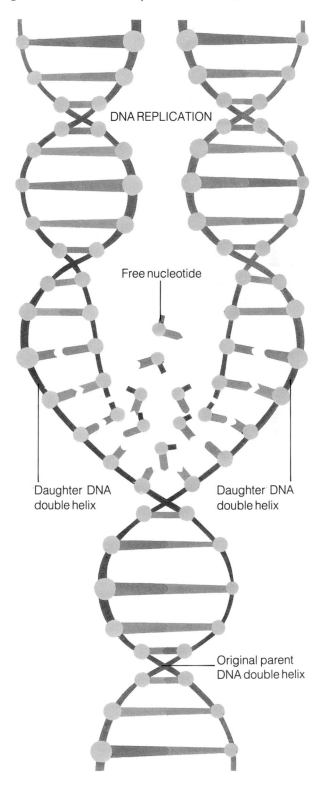

DNA REPLICATION

Free nucleotide

Daughter DNA double helix

Daughter DNA double helix

Original parent DNA double helix

Mitosis—the method by which body cells replicate—has several stages, which are shown in the photomicrographs below (at a magnification of about 450 times). Mitosis is preceded by interphase (top left), *then begins with prophase* (top middle); *this is followed by metaphase* (top right), *anaphase* (bottom left), *and telophase* (bottom middle). *After telophase the cells revert to their interphase* condition (bottom right). *The principal events of mitosis illustrated in the diagram on the opposite page show the important aspects more clearly than is possible using photomicrographs alone.*

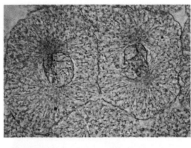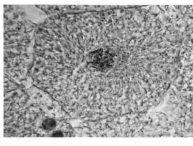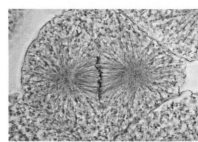

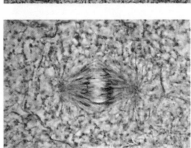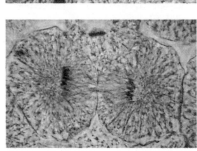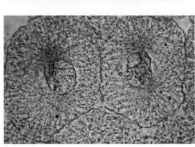

scientists can study chromosomal abnormalities which are linked to certain disorders.

During the next stage, anaphase, the chromosome pairs that are on the cell equator separate and retract toward the ends of the cell along the spindle fibers as if being hauled in.

Finally, in telophase the process of prophase is reversed. The chromosomes disappear, the two new masses of nuclear material regain their interphase appearance, and new nuclear membranes form. The spindle fibers dissolve and the cell cytoplasm divides along its equator, so forming two new daughter cells.

The process of mitosis as observed in the laboratory takes just under one hour; the preceding DNA replication during interphase may take between six and eight hours.

The Aging Cell

Cellular aging and death are as much a part of normal development as is cellular reproduction. Cell death is viewed by biologists as an inevitable event that contributes as much as birth. This idea is not new; the cycle of life was understood by ancient philosophers and observers. Marcus Manilius, a Roman astronomer of the first century B.C. aptly noted, "We begin to die at birth; the end flows from the beginning." But the beginning of death starts even before birth.

As the fetus grows in the womb, some of its cells are dying. This seeming anomaly is not only normal but is actually necessary so that the developing body can discard embryonic organs and appendages such as tails and gills. In mature organisms cell death is necessary for continuing viability. Worn and "tired" cells must be replaced with fresh cells, capable of maintaining functional capacity and reproduction.

Cellular organelles called lysosomes are involved in the dissolution or degradation of old or nonfunctional cells. The lysosomes function in two different ways to accomplish cell destruction. First, the lysosome's outer membrane may break down, so releasing lysosomal enzymes into the cell and consequently destroying cytoplasm and cellular organelles. This natural process is used when the cell is no longer needed by the body to accomplish a particular structural or functional purpose. In other instances the lysosome engulfs part of the cell, forming what is known as an autophagic vacuole.

The consumption of part of the cell does not always herald cell death. The cell uses lysosomal actions to eliminate unneeded parts of itself. Strange as the lysosomal "suicide bags" may seem, it is clear that they are nothing more than part of the grand design of life and death that keeps organisms functioning at their peak.

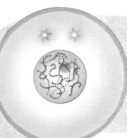

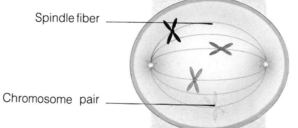

Centriole

Nucleus

Chromosome

8 Interphase: cells separate; DNA replicates

1 Early prophase:DNA condenses to form chromosomes; centrioles move to opposite ends of the cell

Spindle fiber

Chromosome pair

7 Late telophase:spindle breaks down; division is completed; new nuclei form; chromosomes uncoil

2 Late prophase:nucleus disappears; spindle forms; chromosomes become visible as paired structures

MITOSIS

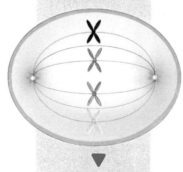

6 Early telophase:cell begins to divide in two by constricting across its equator

3 Metaphase:chromosome pairs line up on the spindle fibers along the equator of the cell

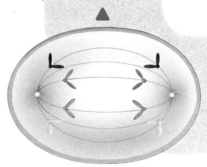

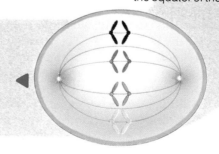

5 Late anaphase:chromosomes reach their destinations at the centrioles

4 Early anaphase:chromosome pairs separate; each half of every pair moves to the opposite end of the cell

*The specialization of cells
plays an essential role — the
tiny embryo must develop all
its different tissues and organs
before it can grow into a
perfect human being.*

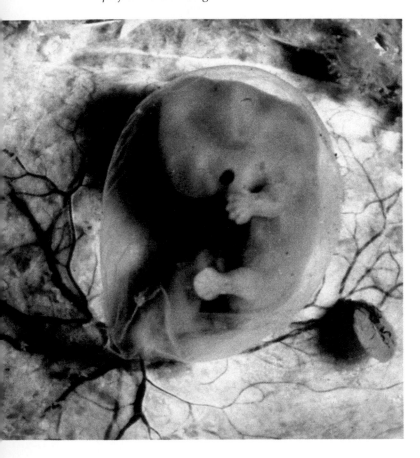

Cellular aging is also a common process, and a natural part of the cell cycle. It was not until the twentieth century, when cell culture techniques made observation of cell generations possible *in vitro*, that scientists began to achieve an understanding of cellular aging. More recently, gerontologists (scientists who study aging) and cell biologists have gained a deeper insight into the mysteries of how cells age, and what the aging process means for the body.

The aging of individual cells seems to be irrelevant, because the waning cell can be replaced by a new cell that is a replica of the old one.

To examine the different aspects of cellular aging, Dr Leonard Hayflick (one of the foremost American specialists on cellular aging) and his colleagues grew a human fibroblast cell line derived from embryonic lung tissue in dishes of culture media where the cells could reproduce freely (a fibroblast is any cell that can develop into connective tissue).

With added nutrients, and at body temperature, the cells covered the surface of the medium in about four days. At this time half the cells were transferred to a new jar. Four days later the cells had again filled the dish. Hayflick observed, however, that after about 50 doublings, cells from human embryonic cultures did not readily divide and cover the available space in the nutrient medium any more. An internal aging clock seemed to tell the cells that their lifespan was nearing its end. This 50 generation limit, within ten generations either way, was subsequently verified in many laboratories around the world.

If the cells really had an aging clock, Hayflick postulated that it might keep track of the number of cell doublings in a person as he or she aged. He therefore tested the doubling ability of cells from individuals of different ages and found that although the number of doublings could not be exactly correlated with the age of the individual from which the cells were derived, there was a pattern. The older the donor, the shorter the *in vitro* lifespan of his or her cells, and cells from older individuals did not usually achieve the maximum number of 50 doublings.

An exception to this pattern was found in victims of progeria and Werner's syndrome, two very rare diseases in which the victims show signs of premature aging. In these cases the number of cell doublings in the laboratory is markedly reduced, despite the young age of the cell donors.

The inability of cells *in vitro* to reproduce indefinitely has also been demonstrated in nonhuman species. Cultured cells from a mink (which normally lives ten years) stop doubling after about 30 generations; those from the Galapagos tortoise, which lives more than 170 years, have an *in vitro* limit of 90 to 125 doublings.

Normal Death Patterns

At the beginning of the twentieth century the development of the technique of growing animal cells outside the body enabled scientists to observe closely cellular behavior and life cycles. Hayflick's experiments in the latter half of the century were possible because of the pioneering work of others, such as Ross Granville Harrison at Yale. In 1907 Harrison grew pieces of tadpole spinal cord, in

A detail from Titian's
Allegory of Prudence
*captures the essence of aging
from a man's callow youth
through mature middle age to
wise but frail old age.*

order to observe how the nerve fiber tips grew. Harrison's work was quickly seized upon by other cell biologists, notably Alexis Carrel, a Frenchman working at the Rockefeller Institute in New York. In 1912 Carrel created a worldwide stir with his work which appeared to show that cells grown *in vitro* could be kept alive forever, hinting at the immortality for which man had so long been searching. Carrel's "immortal" cell line was derived from the embryonic heart cells of chickens. For 34 years the cell line was maintained, outliving even Dr Carrel himself. Newspapers of the time misinterpreted his experiment, reporting that a gigantic live heart was growing forever in Carrel's laboratory.

But in 1961 Hayflick and Paul Moorhead, during the course of experiments using cultured cells, found that they could not maintain cell lines indefinitely. What had been thought to be careless laboratory technique was in fact the normal aging and death patterns that cells exhibit *in vitro*. Despite efforts to maintain cell lines as Carrel had done, the population doublings ceased after 50 or so times in human embryonic cells, and after an even shorter time with the cells from chick hearts. As laboratory after laboratory, using carefully controlled techniques, confirmed Hayflick and Moorhead's findings, the theories of cell mortality that had ruled biology for half a century began to fail.

Later, a number of scientists concluded that Carrel's techniques were the ones that had faults. It is now believed that the nutrient medium he used to feed the chick heart cells was contaminated by fresh embryonic cells that continually added new stock to the line, allowing it to grow and thrive, seemingly forever.

Not all scientists agree, however, that aging of the whole body is ultimately determined by the cell. For those who do ascribe to the cellular control theory, the cell nucleus (which harbors the DNA)

A scientist labels tissue culture dishes prior to using them to grow cells. Tissue culturing enables scientists to study living cells outside their natural environment.

controls the number of generations the cell or cell line may exhibit. To demonstrate nuclear control, Hayflick inserted the nuclei of old cells into young cytoplasm and inserted young nuclei into cytoplasm that had gone through many replications. He found that regardless of the age of the cytoplasm, the cell reproduced itself the number of times that would be expected for the age of the nucleus.

The gerontologists who disagree with Hayflick believe that aging is a whole body function, controlled by the organs and hormone activities that are ultimately responsible for the breakdown of critical body maintenance systems. In a study of death certificates and autopsy data, W. Donner Denckla of the Roche Institute found that humans succumb primarily to the deterioration of two body systems: the cardiovascular and immune systems. Heart and vascular diseases are the most common cause of death in adult Americans.

The thyroid gland is critically important to both cardiovascular status and to growth because it produces hormones that control rate of heartbeat, metabolic rate and the onset of puberty. The thyroid itself is regulated by a hormone (called thyroid stimulating hormone, or TSH) released by the pituitary, a small gland at the base of the brain that secretes a whole host of hormones regulating body processes. Denckla theorizes that interruption of thyroid hormone activity may be responsible for the rate of aging in the body. He further believes that the interruption is caused by a special chemical hormone factor released by the pituitary. As yet, however, no such hormone has been identified or isolated.

The Free Radicals Theory

From the persistent and abiding search for the key to longevity's door has come another promising line of investigation. It is the free radicals theory. Free radicals are common chemical by-products of cellular metabolism. They are capable of bonding to lipid molecules that play an important role in the regulation of cell membrane activities. Free radicals are also capable of damaging DNA (or indeed any other molecule). Usually, free radical interference with normal function is prevented by enzymes that act as anti-oxidants and bond with free radicals before they disrupt the cells. An enzyme called superoxide dismutase (or SOD) is believed to slow the aging process by interfering with the free radicals in this way.

Proponents of the free radicals theory, such as Denham Harman of the University of Nebraska and Roy Walford of UCLA, believe that dietary supplements of non-enzyme anti-oxidants could also prevent the disruption caused by free radicals. Vitamins C and E, the mineral selenium and synthetic food additives such as BHA and BHT are known to have anti-oxidant properties.

Harman was able to demonstrate that the immune response could be enhanced if vitamin E is added to a mouse's diet. He also showed increases in survival rates and delays to the onset of tumors when diets were enhanced by anti-oxidants. Although these laboratory results are encouraging, there is no evidence that anti-oxidants can prolong human life; indeed, some may even be harmful.

Ross Harrison

Pioneer of Tissue Culture

Ross Granville Harrison made important discoveries concerning the embryological development of nerve fibers, and it was in the course of this work that he pioneered the technique of tissue culturing. He wrote in 1937 "The reference of developmental processes to the cell was the most important step ever taken in embryology." Known for his dedication and commitment, Harrison considered honors less important than the pursuit and maintenance of high standards in scientific research.

He was born in January 1870 at Germantown, Pennsylvania, the son of an engineer. He studied zoology at Johns Hopkins University, Baltimore, and at the University of Bonn. Shortly after graduating in 1889, he became Associate Professor of Anatomy at Johns Hopkins University. In 1907 he moved to Yale University, where he was Professor of Anatomy until 1927 and then Professor of Biology until his retirement from teaching in 1938. Subsequently he served as Chairman of the National Research Council until 1946.

Harrison was primarily interested in the development of nerve fibers—a controversial subject at the turn of this century—and he made

several important contributions to this field. In order to observe nerve fibers actually growing and moving, he devised the technique of tissue culturing. He placed portions of nerve tissue in drops of nutrient fluid on the underside of the coverslip of a special microscope slide. Initially he experienced difficulty with this method because he did not realize that the cells needed a solid surface for movement. But he persevered with the technique and, by using a denser nutrient medium, made the first successful tissue culture in 1907.

The results of Harrison's work soon became widely known throughout the scientific community, and the technique of tissue culturing made a great impact on biological and medical research. Culturing made it possible—for the first time—to observe, grow and experiment on living cells outside their normal environment.

Harrison himself did not further develop tissue culturing to any great extent but other scientists seized on it, improving the technique and also using it to make new vaccines (such as the polio vaccine), to test drugs, and to study cancer.

Harrison continued his embryological investigations, which included studies of the effects of transplanting embryonic tissue from one part of an embryo to another. In one experiment, for example, he found that embryonic tissue (from an amphibian) that would have formed a left limb actually developed into a right limb when transplanted to the right side of the embryo. This and other findings helped to establish rules for determining symmetry in embryonic development.

Harrison died in 1959 at New Haven, Connecticut, having provided modern science with the basis of one of its most valuable and widely used techniques, tissue culturing.

Those who study aging often look closely at the aging body to see what changes in cellular function might give clues to the aging process. From the cell they are often able to evaluate implications for the entire body. But some cellular aging phenomena remain a mystery. It has often been shown, for example, that older cells accumulate the pigment lipofuscin. Granules of lipofuscin congregate in the cytoplasm of skeletal, cardiac muscle, liver, prostate, adrenal, ovarian, nerve and other cells. The rate of accumulation is proportional to the age of the animal. This phenomenon is not limited to higher organisms. In some invertebrates, such as roundworms, the age of the animal can be determined by the amount of pigment present in the cells. The origin of and reason for lipofuscin accumulated in older cells is the subject of investigation which is still continuing.

A Clue to Senility

The absence of one particular substance in some aging cells has been shown to have a marked effect on physiological processes. Acetylcholine, a chemical necessary for nerve impulse transmissions in the brain and most of the rest of the nervous system is in short supply in one part of the brain in persons with senile dementia. Symptoms of senility strike about three million Americans. When the number of older persons in the population increases, as is the current pattern, the number of cases of senile dementia also increases. This condition is marked by memory loss and learning deficits. In addition to the enormous health care cost burden, senility in both its mild and severe manifestations threatens the well-being of those caught in its web of confusion, and that of their families, who may well suffer from considerable distress.

The finding of decreased acetylcholine in senile brains comes as a significant step toward possible treatment. Although not all studies with senile patients report increases in memory function when cholinergic drugs (which increase acetylcholine levels) are administered, some gerontologists feel that they may be closer to an understanding of this incapacitating malady.

In response to the special needs of the aging members of the population, the National Institutes of Health in 1940 founded a gerontology unit in the division of physiology. Later this small research facility became the National Institute on Aging, or NIA, and the Gerontology Research Center, or GRC. The GRC is a Baltimore facility where the long-term physiology of aging in individuals is investigated by medical evaluation teams, sometimes for as long as 20 years in the same person. The NIA/GRC is the Western Hemisphere's largest facility devoted solely to the study of aging. The goal of these institutions is "to better the lives of the aged, now and in the future, by improving the social and economic conditions of *all* the aged to the point where they can thrive, not just survive." The accent is on improving the quality of life.

This noble ideal goes far beyond the simple but long-sought goal of extending man's lifespan beyond the biblical three score and ten years. Some populations seem to have found the secret of longevity and continued vitality in older individuals. The extensively studied mountain people of Vilcabamba in Ecuador and of the Caucasus in southern USSR appeared to defy the normal laws that govern the human lifespan. These people lived, it seemed, full and active lives well into their second centuries. Recent studies indicate, however, that many of these people tend to exaggerate their ages, regularly adding a few years beginning at about age 70.

If the human lifespan could be extended for decades or forever, would life be better? Certainly it would not if fraught with physical frailty and mental degeneration. The search for the fountain of youth continues, but now much of the emphasis is on enhancing life. For as Leonardo da Vinci observed nearly five centuries ago, "Life well spent is long."

Goya's portrayal of two haggard, decrepit old women overlooked by an angel caricatures the decay of body and mind—and the imminence of death—that can accompany old age.

The battle between the body's immune system and cancer. The large tumor cell at the top is being attacked by white blood cells (much as the one at the bottom left).

Cell Growth Unchecked

The brain of the human fetus adds thousands of cells to its growing bulk every minute. Yet this remarkable feat has the precise control necessary to ensure that each cell is specialized for its particular function and is added at the right time to the appropriate area of the brain, and that no new cells are added when the organ is complete. This neat and ordered growth of cells into tissues and organs represents the normal developmental pathway. When cellular reproduction breaks that mooring that holds it firmly in proper growth channels, the result is the abnormal proliferation of cells called cancer.

Cancer is a malady shared by man, other mammals, fruit flies, trees and all manner of plants and animals in between. It occurs most often in aged members of the population, but is also found in the young and sometimes develops before birth.

Cancer cells are distinguished from normal cells by lack of control over cell division. If a single cancer cell with a 24-hour generation time is allowed to reproduce freely, without enough nutrients to sustain unlimited growth, at the end of eight weeks the cancerous mass (or neoplasm) would weigh more than 35 pounds and would still be doubling its mass every day. In practice, however, neoplasms usually run out of nutrients before they grow so large.

Cancer is a greatly feared disease, and rightly so, for it is extremely common in our population. Current estimates show that one in four Americans will contract cancer in their lifetime, and one in seven will die from the disease. Furthermore, cancer incidence rates are increasing in general for males, and in specific sites such as the lung in females. This phenomenon is due to a number of factors: a longer-lived population, better detection methods, and an increase in the number of cancer-causing agents (carcinogens) in our everyday environment. When the environment is broadly defined to include personal practices such as cigarette smoking, 80 percent of cancers are thought to be environmentally caused.

Fortunately, treatment and prevention practices are lessening the scourge of cancer, in terms of both suffering and death. There has been an average 18 percent improvement in the survival of cancer patients beyond five years, compared to the survival rate in 1969.

What we call cancer is really a group of diseases, all characterized by uncontrolled cellular reproduction. And just as there are many types of cancer, there are many causes of cancer. For years researchers sought an elusive cancer virus, hoping to eradicate cancer in much the same way that polio and tuberculosis had been quelled. But the cancer virus still eludes researchers, and most now believe that viruses are only a small part of the broad cancer causation problem.

Determining the cause of a particular person's cancer is made even more difficult by the fact that cancer induction usually requires a latency of years or decades between the initial cellular changes and visible onset of disease. Consequently, in most cases it is impossible to pinpoint the exact cause of a specific individual's cancer. In several notable instances, however, the repeated occurrence of a particular type of cancer in individuals has given epidemiologists (those who study disease patterns)

Smoking was once fashionable — as this cartoon of 1827 shows — but has become much less popular with the realization that it can cause cancer, and also respiratory diseases such as bronchitis and emphysema.

PUFF, PUFF, IT IS AN AGE OF PUFFING, PUFF, PUFF, PUFF,

clues to the nature of cancer causation. The common thread might be a job, a habit, or even membership of the same family.

The Viral Factor

Viruses may not be the single causative factor of cancer, but they do seem to play a role in the disease, as research has shown. More than 100 viruses are now known to cause cancer in animals. The first was discovered by the American pathologist Peyton Rous, who in 1910 showed viral infection to be the cause of tumors in chickens. But because his fellow scientists did not accept his work, Rous abandoned it. More than 50 years later the actual importance of his line of investigation became apparent when the Rous Sarcoma Virus was shown to transform normal cells into cancerous ones. At the age of 85, Rous' pioneering work was recognized with the award of the 1966 Nobel Prize for Physiology or Medicine.

A virus does not give rise to cancer in the way a common cold can give rise to a runny nose and other cold symptoms. It is thought that cancer-causing viruses become integrated in the genes and produce cancer when they are activated. Observable changes in chromosomes have been known to be associated with specific animal tumors since 1960. By the mid-1980s about a dozen genes had been related to human cancers of the bladder, lung, breast, blood and colon. The genes found in cancerous cells are actually native to the body's normal cells, where they once existed.

The puzzle is how normal genes become cancer-producing genes, or oncogenes (from the Greek *onkos*, meaning mass). Oncogenes were discovered in viruses before they were found in human cells and recently these viral oncogenes have been shown to be nearly identical to those found in human cancer cells. The viral oncogenes are, in fact, human genes that have been pirated away by the virus and altered slightly. They are capable of being reintroduced into the host's genetic material during viral infection. There, the viral oncogene activates the host cell to initiate the rest of the

Some forms of cancer can be caused by viruses. The Epstein-Barr virus shown below, for example, is associated with nasopharyngeal cancer and with the rare condition Burkitt's lymphoma.

The caption to this 1892 advertisement for Pears soap is particularly apt because cancer of the scrotum was once common in chimney sweeps who did not wash away all of the carcinogenic soot.

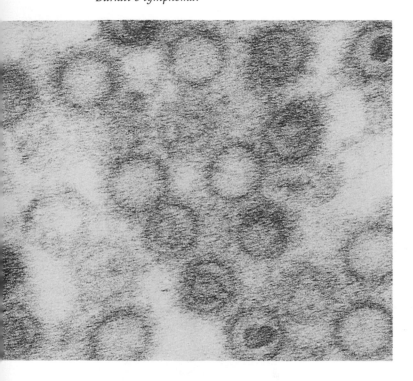

mysterious process that eventually leads to cancer.

Carcinogens, substances that can cause the transformation of normal cells to cancer cells, may function by interrupting the structure or function of the body's normal genes, acting as a switch to turn normal genes into oncogenes. There are many different types of carcinogens, but most authorities agree that they are physical or chemical environmental elements of daily life, such as the free radicals contained in tobacco smoke.

The Environmental Factor

One of the first descriptions of environmentally caused cancers is attributed to Percival Pott, a British surgeon who, in 1775, correlated the scrotal cancer common in chimney sweeps with the failure to remove soot from the folds of the skin during bathing. Since that time the workplace has become a battleground on which numerous wars against environmental carcinogens have been fought. At the beginning of the twentieth century when radium-painted watch dials came into vogue, young women were commonly employed to cover the delicate numerals with radioactive paint. To achieve an accurate brush stroke, they licked the tip

of the brush, ingesting minute quantities of radium as they worked. Years later these women developed cancers of the bone, often in the jaw. Radium is an even more potent carcinogen than the hydrocarbon soot that affected chimney sweeps.

Numerous other substances are carcinogenic: asbestos, coal tar, and aniline dyes are among the commonly mentioned occupational causes of cancer. Each has a common site of action, or a few specific sites that make the cause and effect relationship easier to identify. Asbestos, for example, causes a rare cancer of the lining of the chest and abdominal cavity; when these cancers are seen, asbestos exposure at some time in the patient's past is virtually certain.

Environmental cancer-causing agents are not limited to the workplace. They are found in naturally occurring plants and in foodstuffs. The mold *Aspergillus flavus* — which grows on grain and nut crops stored under hot and humid conditions — produces aflatoxins, potent liver carcinogens. Tobacco products are among the best known and most widely studied natural carcinogens; they can cause cancers of the lung, mouth, larynx, bladder and pancreas.

In addition to naturally occurring chemical carcinogens, a number of synthetic materials — ranging from pesticides and drugs to hair dye components and flameproofing chemicals used in children's pajamas — have been shown to induce tumors when tested in laboratory animals. The similarity of cellular function in all mammalian species has enabled scientists to test many chemicals for carcinogenicity in animals, making it possible to assess how the substance might affect humans. In addition, the universal structure of DNA throughout the living world has fostered the development of new tests that use bacteria, fruit flies and other nonmammalian species to predict carcinogenic activity.

Radiation

Not all cancer-causing agents are chemicals. Radiation, a physical phenomenon, is known to cause skin cancer, and also some forms of leukemia. Sources of radiation include X rays, fallout from atomic explosions, radioactive substances such as radium, and naturally occurring

Good morning! Have you used Pears' soap?

rays and particles that bombard the Earth daily from outer space. And sunlight contains ultraviolet radiation, which may cause skin cancer. Before the perils of radiation exposure were well known, X-ray therapy was prescribed as the cure-all for diseases ranging from asthma to acne; and X rays were even used to check how well shoes fitted! Unfortunately, many patients treated with X-ray therapy developed cancer from their previous exposure.

The exact mechanism by which physical carcinogens cause cancer remains unknown, but it is clear that most, if not all, act to alter the cell's regulatory machinery by damaging the DNA in the nucleus. Any change to the DNA will be passed on to subsequent generations of cells. But not all such changes (called mutations) are reflected in the observable physical characteristics of subsequent generations and not all mutations are expressed as cancer. In fact, the great majority of them are repaired by the cell, or are insignificant and lead to no obvious change, or cause cell death.

Not all carcinogens can initiate the cellular events that lead to cancer induction in the form in which they are first encountered by the cell. Some must act in concert with other carcinogens, called co-carcinogens. Others remain dormant unless activated by undergoing chemical changes in the body. Still others must await the presence of promoter chemicals which alone are not carcinogenic. Most chemical carcinogenesis involves at least two sequential steps.

The Nature of the Disease

Human cancers are diverse, and are often classified according to their site of origin. The major classifications include the carcinomas, which originate in the lining of body organs and surfaces; the sarcomas, originating in the body's connective tissue; the leukemias, which develop in the blood and bone marrow; and the lymphomas derived from the lymphatic system. Classifications are made more specific by adding the name of the organ or tissue in which the tumor originated. For example, adenocarcinoma indicates that the carcinoma began in glandular tissue.

Tumors may be divided into two major classes, depending on whether or not they dislodge from the original site of growth and disperse cells capable of reproduction throughout the body. Benign tumors are those that remain in one place,

Radiation can be a potent carcinogen, particularly "hard" radiation from atomic bomb tests (top). Overexposure to sunlight (bottom), "soft" ultraviolet radiation, can cause skin cancer.

whereas malignant tumors are those that are capable of sending armies of invasive cells to distant sites in the body. The movement of malignant cells from the original tumor site is called metastasis. Cells that have metastasized are capable of establishing new colonies, and it is usually these secondary tumors, rather than the original growth, that are the cause of death in cancer patients.

But not all malignant cells that metastasize survive the journey through the blood or lymph systems; many are destroyed before their ominous mission is accomplished. And although metastasizing cells travel throughout the body, they "stick" and form secondary tumors only in specific organs.

For instance, primary tumors in the lungs tend to give rise to secondary tumors in the brain and adrenal glands. It has been suggested that specific types of tumor cells rely on finding compatible tissues for further growth. Once the cell reaches fertile new ground, it must stake its claim by securing all of the elements vital to its survival. Like normal cells, tumor cells require a supply of nutrients from the bloodstream; the malignant cell

Cancer can spread through the body by metastasis, as illustrated in the diagram below. A primary tumor (shown in red) gradually increases in size (A and B) and begins to shed cancerous cells (C). These cells then move — via the bloodstream or lymphatic system — to other parts of the body, where they establish secondary tumors (D). Often it is the secondary tumors (shown in purple) that cause death.

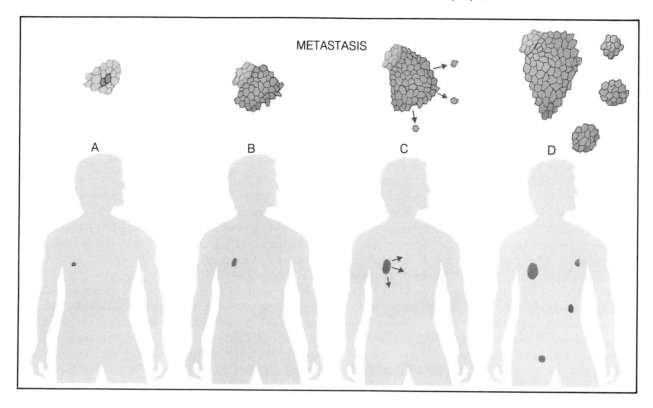

METASTASIS

A   B   C   D

cannot grow into a tumorous mass unless it is well fed. Tumor cells can release a chemical substance known as tumor angiogenesis factor (TAF) that causes nearby blood vessels to proliferate, so that the tumor can lay claim to its own nutrient supply.

By suspending tumor cells in the liquid anterior of a rabbit's eye, Judah Folkman of the Children's Hospital Medical Center in Boston found that blood vessels in the nearby iris began to grow profusely and reach toward the tumor, in response to TAF released by the tumor cells.

A New Hope

In a speech in 1908 a researcher referred to finding a specific drug that would provide an easy cure for cancer. Today, however, we know there is no magic cure. Nevertheless, considerable progress has been made in the fight against the disease — by ever-improving uses of traditional treatments, and by the application of new research areas that can detect cancer sooner, lessen the suffering of its victims and restore them to health.

One of the most promising cancer detection and cure prospects involves the use of the cancer cell itself. By combining a tumor cell with an antibody-producing spleen cell, scientists are now able to construct a cellular factory called a hybridoma that can produce highly specific monoclonal antibodies. Furthermore, it can produce limitless quantities of these antibodies because tumor cells are not subject to the normal 50 generation limit; in effect, the tumor cell confers immortality on the hybridoma. The antibodies are capable of tracking and adhering to target proteins, such as cancer cell surfaces. In this way, a monoclonal antibody labeled with a radioactive tracer can search out cancer cells before the tumor is large enough to be seen by conventional means such as X-ray or CAT scans (X-ray scans that show slices of the body). In laboratory tests hybridoma-produced antibodies successfully destroyed cancer cells living in culture and it is hoped that the hybridoma technique can be developed sufficiently to enable the antibodies to be used as ''guided missiles'' armed against specific cancer cells but harmless to the functioning of normal cells in the body.

Most cancers are currently treated by one or a combination of three methods: surgery; radiation (radiotherapy) usually using X rays; or drugs (chemotherapy). Surgery is used first on more than one third of cancers to excise the tumor mass or the cancer-riddled organ. It is highly successful with benign tumors, and with malignant tumors that have not metastasized.

Radiation works by physically destroying the cancer cells. It is not selective, however, and healthy cells are inevitably damaged during the radiotherapy. In a combination therapy, radiation may be used to reduce the population of remaining cancer cells that were not surgically removed. In this way, it is useful in managing malignancies that have spread. The National Cancer Institute is developing neutron generators that will be more effective than conventional X rays. It is also working on drugs that may inhibit the adverse effects of radiation on healthy cells.

Chemotherapy reaches deep into the hidden recesses of the body, to search out and destroy cancer cells that cannot be eliminated successfully by surgery or radiotherapy. Anticancer drugs have the disadvantage of also destroying other fast-growing cells in the body, such as hair follicles (which results in hair loss during treatment), and of producing other adverse effects — nausea, for example. More than two dozen cancer-combating drugs are available, and with these the ability to successfully treat cancer improves. There was, for instance, no cure for Hodgkin's disease in 1963 but now this lymphoma is treatable; indeed, the death rate from Hodgkin's disease has diminished and today more than three-quarters of patients are cured.

Cancer's toll continues to fall, not because of a single mystical formula, but because concerted and highly specific research suggests new ways to prevent, detect and cure.

Use of the body's own defense system, the immune system, is gaining momentum as an alternative to traditional cancer treatments. In theory, if the immune reaction against the cancer cells could be bolstered, the body might be able to eliminate cancer cells in much the same way it does other foreign substances such as viruses and bacteria. Immunotherapy involving a bacterial derivative that encourages the body's own immune system has shown promise when administered after chemotherapy in cases of leukemia, Hodgkin's disease, breast and other cancers.

Another line of research has been into finding proteins already existing in the body naturally, although possibly in minute quantity, that may act as anti-cancer agents. One such protein is actually produced by all cells in the body: interferon. Cells produce interferon after viral infection, and although it has no direct effect on the virus, it protects other cells from infection. One effect of its use in therapy, however, is to cause all the symptoms of influenza (and some physicians think that interferon may accordingly be the real cause of influenza symptoms in patients suffering from that disorder.)

Interferon, or IF as it is known for short, is now being evaluated for its therapeutic activity against cancer as well as against viral infections. To date, clinical trials have shown some anti-cancer activity — such as a reduction in tumor size — with the substance, but these trials are in the earliest stages and investigators emphasize the preliminary nature of their work. In fact, some stress that other widely used chemotherapies seem to be equally effective. Interferon may not turn out to be the panacea for cancer, but it may help to prolong lives.

In 1975 American scientists first described a similar protein now known as tumor necrosis factor. This is a protein produced by macrophages, which under laboratory conditions with specially cultured tissues has been shown to kill certain cancer cells in mice, and to retard the development of other cancers. It is hoped that tests will prove that human cancers may be successfully treated with tumor necrosis factor, although it would probably be more realistic to expect much the same results as with interferon. Initially the factor was in short supply — indeed, so short that many physicians simply did not believe it existed — but through genetic engineering, the factor is (at least in the United States) available for experiment.

Another form of anti-cancer agent presently only suspected to exist is made up by the chalones. What is called the chalone concept suggests that there are substances specific to each type of body tissue which signal to the cells when they have reached

Tumor angiogenesis factor (TAF) is released by tumor cells to increase their supply of blood. The diagram below shows the normal condition, before a tumor has developed. When cells become cancerous (shown in red), they release TAF (middle diagram), which causes nearby blood vessels to proliferate (right). Because of the increased blood supply, the tumor receives more nutrients and grows rapidly.

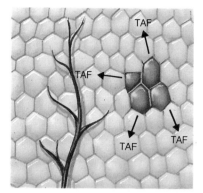

maturity, so they undergo no further division. Skin cells in particular are reckoned to have a chalone so that they die and slough off in enormous numbers daily. If they can be shown to exist, chalones would probably be useful as anti-cancer agents.

The HeLa Cell

In 1951 a 31-year-old Baltimore woman died of cervical cancer that had been diagnosed only eight months before. Her name was Henrietta Lacks and her death was remarkable in one particular way: the cells that were taken from her malignant tumor are still alive today. HeLa cells, as they are known in laboratories around the world, were cultured in the Johns Hopkins University laboratory where they had first been taken for diagnosis and soon became popular tissue cultures for the study of cancer and cell growth. They were even used during the development of a vaccine against polio.

The cells of Henrietta Lacks — and cells of all other cancers — demonstrate the vigor and lack of growth control that distinguish cancer cells from normal cells. Cancer cell populations defy the laws that govern the reproduction and growth of normal cells and, under the right conditions, can live forever; not only do they not know when to stop dividing, but they also do not know when to die. Their life cycles are uncannily like those of normal cells. The DNA is copied, the cell divides, the cell ages and dies. But in cancer cells there is no clock that times the overall life of the cell line; they appear to be able to go on and on multiplying. Here, then, is the ultimate irony: cancer, perhaps the most feared of diseases, makes cells immortal, and causes them to disregard the carefully evolved calendar that normally determines a cell's life.

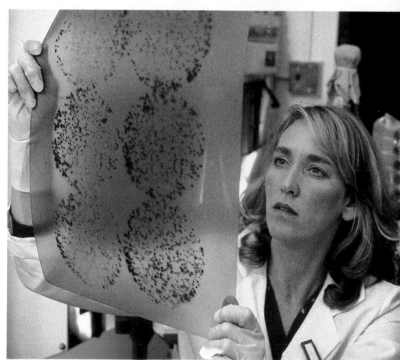

As part of a cancer research program a scientist examines an autoradiograph on which the black spots represent colonies of viruses that have been engineered to contain potentially cancer-causing genes.

Chapter 5

Attackers and Destroyers

Samael, the Angel of Death, is a figure that occurs often in German imagery of the Devil and Death. Throughout medieval Europe, these skeletal, scythe-bearing figures were frequently associated with diseases reaping life in a great human harvest, and as a punishment for wrongdoing rather than as a natural cause through the breakdown of physical processes.

an has forever cast some of his deepest dreads and most fundamental fears in images of pestilence and disease. Centuries before disease was found to spread from natural sources, it was believed to be morally, as well as literally, contagious, a sign of corruption and a means of divine retribution. Arousing the menace and mystery of death itself, disease was masked in myth, magic and metaphor. In Dante's *Inferno*, for example, the lowest circles of Hell are peopled with the fraudulent whose sin — itself "a canker to every conscience" — God loathes above all and punishes with disease. One of these circles resembles "the hospitals of pestilence," crammed with sinners, themselves corrupted just as they corrupted others.

The ancients suspected that disease passed from person to person, but supposed that spirits, divine and demonic, shepherded its malevolent meanderings. In the Old Testament, the Mosaic law treats leprosy as both a disease and a disgrace. In near-clinical language, the law of the leper begins by describing the signs of the disease. The priest, suspecting leprosy, isolates and confines the patient for two weeks. Every seven days he checks the symptoms. If, at the end of a fortnight, his suspicions of the disease are confirmed, he declares the patient a leper. As long as "the plague shall be in him," reads the law, "he shall dwell alone: without the camp shall his habitation be." The leper remains in isolation until the priest declares him cleansed of the disease and he ceremonially atones for his sins.

Combining spiritual absolution with physical purification, the priest then ensures that the leper's clothes, belongings and home, any of which might carry the disease, are destroyed or purged. In this way, making use of a combination of medicine and mysticism, the law of the leper hinted at the nature of contagion while tracing disease itself to supernatural sources.

The Classical Theories

In contrast, the classical physicians of Greece and Rome overlooked contagion but attributed disease to natural causes. Epidemics were linked to such natural phenomena as eclipses, comets, earthquakes and floods, but most of all to seasonal and climatic changes in the atmosphere. The *Corpus Hippocraticum*, a series of medical treatises attributed to Hippocrates, the foremost physician of the fifth century B.C. (but written by many), proposed that disease was caused by "miasma," or vapors which fouled the air. The miasma theory of disease was complemented by the Hippocratic doctrine that the body housed four humors — blood, phlegm, black bile (melancholy) and yellow bile (choler). These humors were influenced by the four elements — earth, air, fire and water. Impure air upset the balance of the humors to bring about disease. This theory, though now long rejected, caught the imagination of writers right up to the nineteenth century. It appears in the works of such authors as Emile Zola who, in his book *Germinal*, criticizes mining conditions, and Thomas Mann, in

whose *The Magic Mountain* the hero is surrounded by disease and death in a tuberculosis hospital.

Some laymen, perhaps less encumbered by medical metaphysics, assumed that disease was spread by contact. In his famous account of the plague of Athens in 430 B.C., Thucydides the historian, himself a surviving victim of the disease, leaves its causes to physicians but stresses the plague's contagious character, observing that death was common among doctors who succored the stricken, and animals feeding on the flesh of the fallen also soon perished.

Roman historians, philosophers and poets also considered disease contagious, although Galen, the Empire's medical authority, clung to Hippocratic teachings. Livy, recalling the plague that struck the armies of Rome and Carthage at Syracuse in 212 B.C., says the disease started with climate but spread by contagion "until those who caught it were either left to die alone or took with them to the grave whoever sat by their bedside and tried to tend them." After counting the toll of anthrax on livestock, Virgil explains that even tanners refused the hides because:

> . . . if the vest they wear
> Red blisters rising on their paps appear,
> And flaming carbuncles, and noisome sweat,
> And clammy dews, that loathsome lice beget;
> Till slow creeping evil eats his way,
> Consumes the parching limbs, and makes
> the life his prey.

Stealing and skulking amid mankind, the causes of widespread disease clearly escaped the speculations of physicians more easily than the imagination of writers.

The Christian View Inspected

While disease, taking alike young and old, rich and poor, good and bad, was just another of the several disasters that sapped the strength and spirit of Rome, its ravages were, to Christians, a revelation of the purposeful hand of God. Cyprian, the Bishop of Carthage, celebrated the plague that raged there in 251 A.D. saying, "How suitable, how necessary it is, that the plague and pestilence, which seems horrible and deadly, searches out the justice of each and every one and examines the mind of the human

The image of thick, impure air as a cause of disease and ill health was originated by the ancient Greeks, and corresponded to a disequilibrium of the four humors contained in the body—yellow bile, black bile, phlegm and blood. It was an image which inspired writers and painters for centuries afterwards, but particularly in the wake of the Industrial Revolution in Europe. At that time, poverty in the rural areas had driven the masses to the cities. There, in cramped, polluted slums, their conditions were exacerbated by the smoke- and soot-clogged air. The result was that waterborne and bronchial diseases were rife.

race." Unlike pagan philosophers, Christian theologians were not shaken for want of rational explanations of epidemics, but instead accepted their horrors and hardships with the determination they inspired. As one epidemic followed another across medieval Europe, the official wisdom of Church and medical authorities was increasingly challenged and found wanting. The random, indiscriminate affliction of disease led some to question whether, as William Langland, the author of the English epic *Piers Plowman*, put it "these pestilences were for pure sin." The Florentine poet, Boccaccio, setting his masterpiece, the *Decameron*, amid the Black Death of 1348, suggests that the plague was less the consequence, and more the cause, of sin for, "In the face of so much affliction and misery, all respect for the laws of God and man had virtually broken down and been extinguished in our city." Guy de Chauliac, the most celebrated of the physicians then tending the Pope at Avignon, said simply "Charity is dead."

For Boccaccio, what made the plague so severe was that "whenever those suffering from it mixed with people who were still unaffected, it would rush upon these with the speed of a fire racing through dry or oily substances that happened to be placed within its reach." He recalls that despite various fears and fantasies aroused by the epidemic, those alive and well "almost without exception ... took a single and very inhuman precaution, namely to avoid or run away from the sick and their belongings." Boccaccio tells of brother abandoning brother, husband leaving wife and parents deserting children. Little wonder a Flemish cleric at Avignon called contagion "the most terrible of all the terrors."

In 1403, some Venetians who suspected, with others, that plague came from the Levant, opened a hospital on an island beyond the city where travelers from the East were confined. Their isolation and its duration — the 40 days and 40 nights of the biblical flood — typified the mixture of hygiene and faith with which disease was countered, and it also left the Italian word *quarantina* (40-day period) as a legacy to the medical vocabulary.

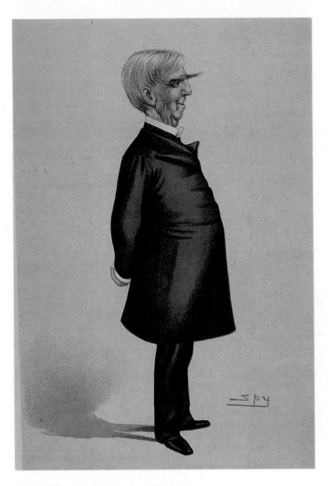

The Germ Theory of Contagion

It was not until the sixteenth century that contagion entered medical literature as an explanation for the spread of disease. By then bubonic plague, and several other epidemic diseases as well — smallpox, typhus, measles, syphilis and "the English sweats" — had been closely observed and carefully described. Even so, the contagious nature of disease became most horribly evident during Europe's conquest of the New World at this time. Europeans introduced diseases previously unknown among the American Indians, such as smallpox, which decimated populations and inflicted pain, disfigurement and death on succeeding generations.

Girolamo Fracastoro, a gentleman scholar and poet of Verona, marshaled this new knowledge into a theory of contagion, presented in a short Latin treatise in 1546. Fracastoro begins by defining contagion as an infection which passes from one individual to another, distinguishing it clearly from the putrefaction of rotting flesh and foodstuffs. He called the agents of contagion *seminaria* (seeds or germs), although he refrains from characterizing them precisely. Fracastoro presumed contagion always occurs by contact but conceded that it may occur through contact with "fomites" — clothing and the like — which harbor germs without themselves being infected. Recognizing differences between diseases, Fracastoro supposed that some attack one organ, others another. He followed the courses of different diseases, noting the varying times between exposure to the disease and the appearance of symptoms. He also pointed to the changes in symptoms that some diseases, particularly syphilis, produce over a period of time.

Like the astronomers who deduced the existence of Neptune from perturbations in the orbit of Uranus, Fracastoro's insights reached beyond the commonplace knowledge at hand and the understanding of his contemporaries. The clues Fracastoro left were soon lost and forgotten but during the three centuries following his death, his finding were gradually discovered afresh until the germ theory of contagion he sketched so boldly was at last confirmed.

In the mid-seventeenth century, with the development of the first microscopes, the Dutch biologist Antonie van Leeuwenhoek explored the world of minute organisms. But despite Leeuwenhoek's discovery of "animalcules," few scientists made the conceptual connection between tiny life forms and disease. An exception, however, was Benjamin Marten, an obscure London physician who offered remarkably prescient views on disease in his *New Theory of Consumptions*, printed in 1720. Consumption, he suspected, was caused by "some certain species of animalcula or wonderfully minute living creatures that by their peculiar shape or disagreeable parts are inimical to our nature." And he guessed that different species of "animalcula" spawned different strains of disease, even arguing that they, not climate, heralded seasonal recurrences of diseases. But Marten, like Fracastoro before him, was ignored by his contemporaries and the germ theory of contagion was disregarded until the late nineteenth century.

More than a century later, in 1842, Oliver Wendell Holmes, an American physician and writer, insisted that puerperal fever, a scourge of childbirth, was contagious, carried from mother to mother by the doctors and midwives who delivered their children. At about the same time, a Hungarian

obstetrician, Ignaz Philipp Semmelweis, pioneered the use of antiseptics for childbirth. Although Semmelweis markedly reduced the incidence of disease and death, his critics not only challenged his procedures but hounded him from his professorship, leaving him to die in crazed anonymity.

In retrospect, the resistance of the medical community to a germ theory of contagion is puzzling. Isolation and quarantine had been practiced for centuries and had been shown to be effective. Leprosy, for instance, had been all but eradicated from Europe. In England and France, the lazar houses, in which lepers were isolated, had long been put to other uses. The miasma theory of disease, though long embedded in the popular and literary imagination, no longer commanded such respect among physicians. As early as 1801 the distinguished French anatomist and pathologist Marie-François-Xavier Bichat proposed that disease be approached as "pathological life," with a cause, career and conclusion of its own. Yet broadminded, educated men who readily accepted a universe ruled by gravity — an unseen force acting in empty space across vast distances — could not

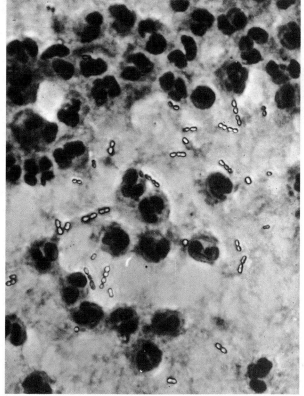

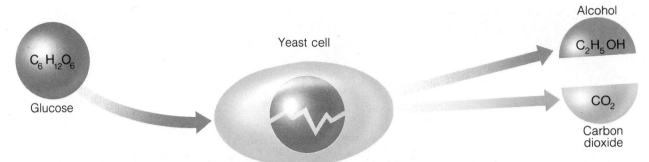

Fermentation was first found by Louis Pasteur to involve yeast cells. When glucose is added to a yeast cell the glucose is broken down. During the process carbon dioxide is given off and alcohol is produced.

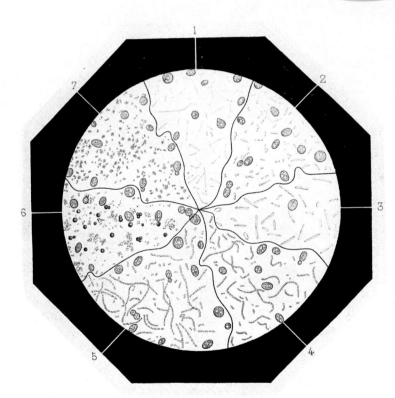

Among the many drawings Pasteur made of his studies of failed fermentation was this one which depicts the organisms found in various substances that had degenerated. The substances are: (1) turned wine; (2) soured milk; (3) butyric ferment; (4) what he calls "ropy" wine; (5) vinegar; and (6) an amorphous deposit. A saprophytic bacterium of the genus Sarcina *is shown in 7.*

countenance the notion that microorganisms carried disease from one person to another.

Although among these skeptics, Jacob Henle, a German anatomist, made a significant contribution to the germ theory in an essay he wrote in 1840 in which he spelt out the conditions a germ theory of contagion must satisfy to be regarded as proved. Before any microorganism could be considered to be a cause of a disease, it must, he insisted, be present in every case of the disease tested. And, when isolated and introduced into a previously healthy person, it must produce the same disease in that person. Before the century was out, a young chemist, Louis Pasteur, and one of Henle's own pupils, Robert Koch, met his challenge and established the theory.

At Lille, where he was appointed professor of chemistry in 1854, Pasteur was encouraged to apply his learning and talent to the problems of local industries, chief among them being the production of alcohol from beet sugar. French chemists of the time taught that fermentation, along with all other changes in organic matter, was strictly a chemical reaction, and overlooked the action of yeasts and other living organisms. Challenging convention, Pasteur demonstrated the role of yeasts in fermentation, showing that alcohol production was directly proportional to the amount of yeast used. Similar studies of the souring of milk and vinegar production enabled him to confirm that microorganisms could bring about changes in organic substances. Since microorganisms were necessary to certain processes, he suspected that alien ones might disrupt or foul them. After proving that bacteria competing with yeasts could spoil wine and beer, Pasteur devised the heating process to kill bacteria that bears his name.

Convinced that microorganisms affected organic

processes, Pasteur turned to yet another ailing industry. Midway through the nineteenth century, disease struck Europe's silkworms, bringing the silk industry to the brink of bankruptcy. A century before, when Italy's silkworms were stricken with disease, Agostino Bassi, an amateur scientist, had shown that they suffered from a contagious fungus infection. But Bassi's discovery, the first to reveal a contagious disease in animals, went unnoticed outside Italy. Knowing nothing of silkworms, Pasteur began his research in 1865 and, five years later, demonstrated that some contagious diseases among silkworms were caused by germs.

In England, at about the same time, the surgeon Joseph Lister had grown concerned at the number of deaths caused by wound infection. He believed that germs existed in wounds, and that others could enter from outside, causing suppuration and putrefaction. His researches led him to discover carbolic acid (now usually known as phenol) as a means of killing germs when applied both to the wounds and impregnated in dressings, and from this to the introduction of the antiseptic principle in surgery. Lister's findings were published in 1867. Like Pasteur's, they predated the proof that germs cause disease.

The Discovery of Germs and their Suppression

Far from centers of learning, the doctor Robert Koch pursued a lively medical practice among the farming families of the flat, sandy, windswept plains of East Prussia. Koch, a learned, curious and ambitious physician, grew frustrated with treating diseases whose mysteries he did not understand. He equipped a laboratory next to his clinic and started searching for the source of disease, beginning with anthrax, which was common among farmers and their stock.

Koch was exceptionally patient and painstaking, and developed methods and techniques to isolate specific bacteria from the tissue and blood of diseased animals. He found that one bacterium was invariably present with anthrax, and this he bred in pure cultures — exclusive colonies free from other microorganisms. Koch then injected bacteria drawn from his pure cultures into healthy laboratory animals, which soon developed anthrax.

In 1876 Koch presented his biography of the

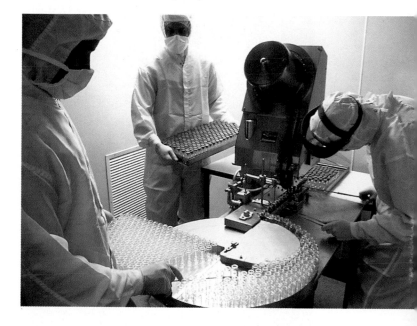

In studying microorganisms it is important to prevent contamination by unwanted species. This is achieved by rigorously maintaining sterile conditions, as in this laboratory.

anthrax bacillus to an audience of prestigious medical scientists. After listening to him for three days, none doubted that the young country doctor had not only confirmed the germ theory of contagion but had also refined the procedures and techniques its application required. Koch formally expressed the theory, together with the procedures required to confirm it, in an article written in 1890. His rules for determining the cause of a particular disease were almost the same as those of Henle: that the microbe must be associated with each instance of the disease; that the microbe must be isolated in a pure culture; that microbes from the pure culture, when introduced into healthy animals, must reproduce the same disease; and that the microbe must be isolated from the second recipient. It was Koch rather than Henle who showed how these requirements could be met, and his rules have since been known as Koch's postulates.

With Koch's findings soon confirmed by those of other scientists, medicine embraced the germ theory of contagion as firmly as it had once rejected it. One after another microorganisms, popularly known as microbes, were identified and classified. Their roles in various biological processes, many important to agriculture and industry (such as baking and brewing), were also explored.

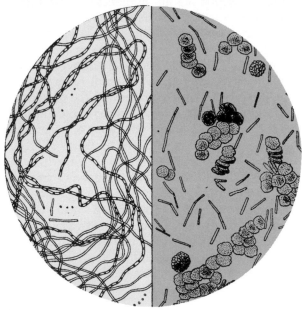

Robert Koch's study of anthrax bacilli in the fluid of an ox's eye revealed their rodlike shape and their rapid reproduction, which culminated in the release of many black spores.

Unidentified Viral Objects

Bacteria were the first microbes found to cause disease in man, and their study — bacteriology — soon led the field in microbiology. But in their study of bacteria, scientists soon picked up signs of other minute, disease-causing organisms.

Since Edward Jenner's remarkable work in the 1790s, people had been protected against smallpox by inoculation with a substance drawn from the lesions of cowpox, a much less dangerous disease. But although he discovered how to control smallpox, Jenner never identified its cause. Nearly a century later, Pasteur, approaching 60, set out to discover why some diseases, like smallpox, conferred everlasting immunity, and to test his hunch that Jenner's procedure, or something like it, might also prevent other such diseases. He began with the assumptions of bacteriology, followed its methods and used its techniques. By breeding germs in laboratory animals, Pasteur found that under certain conditions the virulence of the germs diminished with succeeding generations. He also found that inoculation with these "attenuated" germs conferred immunity — a procedure called vaccination, from *vacca*, the Latin for cow. He developed vaccines against several diseases and in 1865 crowned his career by developing a vaccine to protect those at risk of infection with rabies. Yet Pasteur, like Jenner, never identified the organism that he turned from the cause of disease into its protection.

Meanwhile, microbiologists were making greater use of elaborate filters, used successfully by Pasteur and Koch, to trap bacteria. In 1884, a young Russian, Dmitrii Ivanowsky, studying tobacco mosaic disease, reported that the agent of the disease passed through his filters to infect healthy plants. Two years later, Martinus Beijerinck, a Dutch plant pathologist, discovered that he could not grow cultures of the agent of tobacco mosaic disease in the usual glass dishes. It reproduced, he observed, only in living tissue.

Until then microbiologists referred to all infectious agents indiscriminately as viruses, a common — even colloquial — word from the Latin term meaning slimy or poisonous fluid. Believing, like Pasteur, that all viruses are microbes, they drew no distinctions between agents of disease. Those that

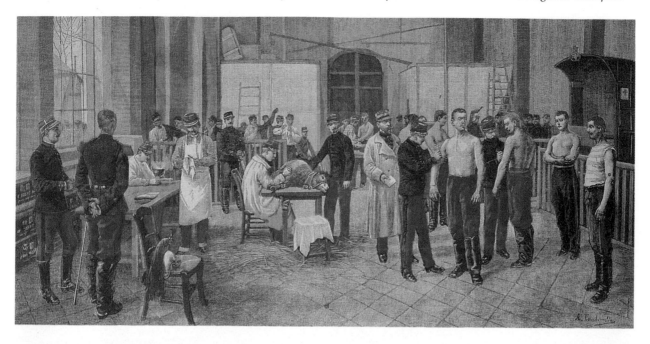

escaped filters were first called "filterable viruses," but, until Beijerinck, were assumed to differ from bacteria only in size.

Venturing an idea years ahead of its time, Beijerinck proposed that a filterable virus differed from other microbes because it reproduced only when "incorporated into the living protoplasm of the cell, into whose reproduction it is . . . passively drawn." With Beijerinck serving as midwife, bacteriology prepared to deliver its first offspring — virology. But not until the 1920s did virologists isolate, observe and analyze the character of viruses, confirming Beijerinck's original suspicion that they were a different sort of organism.

Microbial Opportunists

With bacteriology in full stride and virology on its feet, the conception of disease soon changed from a struggle for the soul to a battle for the body. Hosts of hostile germs relentlessly stormed and breached the body's ramparts as military metaphors replaced spiritual ones in man's imagery of disease. Scientists, transformed into "microbe hunters," stalked the enemy, defeating germ upon germ in skirmish after skirmish as they waged the endless war on disease.

Behind this image of hordes of "bad" micro-organisms being hunted down, the true picture is somewhat different. In fact, the majority of microorganisms play a beneficial role in man's life. Often smaller than the cells they strike, the disease-causing bacteria and viruses are renegades. They are the wayward few among myriads of microbes thriving in every corner of the world, most of which not only cause man no harm, but actually perform tasks vital to his life and growth.

For all its scientific power, the germ theory of contagion paints a distorted picture of disease. Although bacteria and viruses bring about disease, they act in concert with the body itself. In different ways, both bacteria and viruses are parasites, sustaining themselves at the expense of their hosts. Whether or not a microbe is pathogenic (capable of producing disease), and the degree of its virulence, depend both on the character of the microbe and the condition of the host.

At the same time, the germ theory brands specific microbes as pathogens, a judgment microbiologists seem more and more reluctant to make. The body plays host to an enormous number of bacteria of numerous different species, its "normal flora." For instance, there are more bacterial cells in the large intestine alone than there are human cells in the entire body. Most bacteria in the body are harmless;

Wars during the nineteenth century seemed almost as abundant as the pathogenic microorganisms they encouraged. Some soldiers wounded during the Crimean War were lucky enough to be cared for by Florence Nightingale, who insisted on the highest standards of cleanliness and nursing possible in those conditions. The unlucky ones often lay on flea- and rat-infested mattresses with filthy bandages and wounds.

indeed, many are beneficial. But some have been found to produce disease in response to changes in the condition of the host. Instead of judging microbes guilty or innocent, microbiologists now prefer to consider all parasitic microbes as opportunists, acknowledging that they are all potentially pathogenic. Also, just as members of the "normal flora" may cause disease, pathogenic microbes may become part of the "normal flora."

The Body's Defense

To survive in a host, parasitic microbes must be able to metabolize, or sustain themselves, and to reproduce their kind in host tissue. At the same time they must be able to resist the defense mechanisms of the host.

The human body protects itself against invading microbes by a combination of preexisting defense mechanisms and acquired immunity. The existing mechanisms consist of mechanical barriers such as the skin and epithelial cells lining the air passages and gastrointestinal tract. The air passages, eyes and digestive systems are also protected by fluid secretions. For example, tears wash away microbes and also contain enzymes, such as lysozyme,

which will kill bacteria. These represent the body's first line of defense.

If bacteria manage to penetrate this first line of defense, then a complex set of reactions known as acute inflammation is triggered off. Phagocytic white blood cells enter damaged tissue from the bloodstream along with proteins from the blood which help immobilize, neutralize and finally kill the invaders. This is the body's second line of defense.

Particularly virulent organisms may escape destruction by acute inflammation. To cope with these invaders the body makes specific antibodies which can bind to hostile microbes and kill them. It takes one to two weeks for the immune response to reach its peak the first time the body struggles with a particular organism. Thereafter the immune system has a type of specific memory which allows it to respond rapidly when the same organism appears again.

Immunity can be stimulated by the use of killed or altered microorganisms which are made into vaccines. Vaccines often provide the first exposure to a disease-causing organism, so that the immune system can respond rapidly to subsequent ex-

Bacteria are classified into three types, depending on their basic shape. Those that are spherical are called cocci, and may occur as separate, single cells, clumped, or joined to form a chain. Bacilli, the rod-shaped bacteria, are also found singly or linked and may have tails or may be covered with flagella (tiny hairs). Long, twisting bacteria are called spirochetes; shorter ones are known as vibrios.

posure. Many of the symptoms of disease, such as pain and swelling are actually manifestations of these defense systems going into action. What we perceive as disease is really a response to tissue damage.

Bacteria and viruses, the most common agents of disease, work very differently from each other. As Tolstoy wrote of families in *Anna Karenina*, "All happy families are alike, but an unhappy family is unhappy after its own fashion," so sound health is the same the world over whereas infection generally takes many varied courses. And, since disease-causing microorganisms behave differently in the laboratory from the way they do in the body, the dynamics of disease are often obscure. Little is known, for instance, about how bacteria elude the body's interior security services. But once on the loose, they cause all sorts of trouble.

Toxins, Enzymes and the Immune System

Some bacteria, though by no means the majority, produce poisons or toxins which take two forms —exotoxins and endotoxins.

Bacteria make exotoxins when they are themselves diseased. Tiny viruses, called bacteriophages, can infect bacterial cells, prompting them to produce exotoxins. For example, the diphtheria bacillus, *Corynebacterium diphtheriae*, and several streptococci (which cause sore throats as well as scarlet and rheumatic fever), produce exotoxins when infected in this way. The bacteria release their exotoxin into the medium around them. The diphtheria bacillus grows in the throat and spills its exotoxin into the bloodstream. *Clostridium botulinum* occasionally releases its exotoxin into the liquid filling canned foodstuffs and causes botulism in those who eat the contents. And *Clostridium tetani*, unable to encroach or thrive on healthy tissue, grows on damaged or dead tissue where it releases the exotoxin that causes tetanus.

The exotoxins responsible for the noxious effects of some bacteria are proteins. Heating toxins or exposing them to acid changes the shape of the proteins so that they are no longer toxic but can still provoke an immune response. Toxoids, as the modified and now harmless toxins are called, are used as vaccines so that the body develops specific antibodies against them. These antibodies can

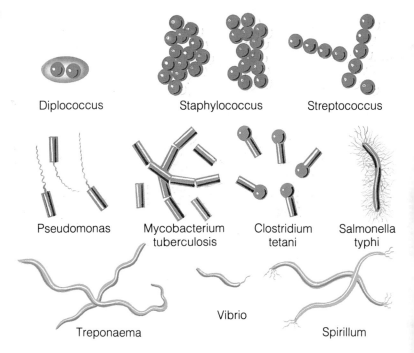

Diplococcus · Staphylococcus · Streptococcus

Pseudomonas · Mycobacterium tuberculosis · Clostridium tetani · Salmonella typhi

Treponaema · Vibrio · Spirillum

combine with the parent toxin and neutralize it when the toxin-producing microorganism causes infection. The best known toxoid, tetanus toxoid, is made from tetanus toxin. It acts to protect against the paralytic effects of tetanus toxin. Occasionally, an individual not immunized with tetanus toxoid in childhood contracts tetanus from a contaminated wound. In this case it is too late to administer the toxoid because, as we have seen, making antibodies is a slow process. Instead, it is necessary to give antibodies from immune donors who have high concentrations of antitoxin to neutralize the poison.

Endotoxins are contained in the cell walls of bacteria and are released when the walls disintegrate. They are complex molecules containing fats, sugars, and proteins, known as lipopolysaccharides. Gonococci and meningococci, the bacteria causing gonorrhea and meningitis, produce endotoxins, as do the enterobacteria, a large family of bacteria, several species of which reside in the intestinal tract. Among the enterobacteria, a number of "opportunists" (which cause trouble only when the body's defenses are down) and occasional pathogens are responsible for a

The Proteus *bacterium* (below) *is a bacillus, a rod-shaped cell covered in many fine hairs. It can make the enzyme phospholipase, which is also an active ingredient in the venom of the spitting cobra.*

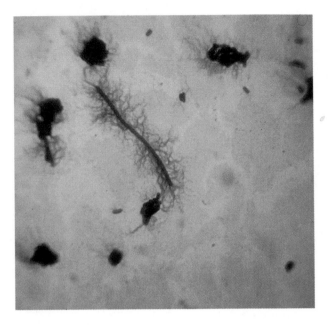

variety of uncomfortable disorders, dysentery and gastroenteritis being the most common.

Endotoxins provoke a vigorous response from the body's defense systems and occasionally the response is so extreme that the host is killed instead of the bacteria. Because endotoxins do not stimulate strong antibodies it has not been possible to make vaccines against endotoxin-bearing bacteria. Instead the body uses a group of proteins (collectively called the complement system) which coat bacteria and help to kill them directly or have them taken up by phagocytes (engulfing cells) in the liver and spleen where they are killed. The complement system can also activate bloodclotting and causes white cells to form clumps. The net effect of overactivation of the complement system by endotoxins is to block small blood vessels in the lungs with aggregated white cells and to form small bloodclots in small arteries elsewhere in the body. Fortunately this potentially life-threatening reaction does not usually happen, and the complement system acts as an efficient defense against these organisms without causing any damage.

Endotoxins also provoke other responses from the body. They stimulate white blood cells to produce a fever-causing substance called interleukin, which causes a "temperature," and provoke the release of white cells from the bone marrow, causing the raised white blood count commonly found during infection. Thus endotoxins are the trigger that releases a wide range of defense systems to bring invading bacteria under control.

Cortisone, a hormone produced by the adrenal glands, possesses the capacity to dampen defensive responses to bacterial infection. In the 1950s, not long after it had been synthesized and had undergone its first clinical trials, cortisone was given to patients stricken with pneumococcal lobar pneumonia and primary atypical, or mycoplasmal, pneumonia. Within hours, the patients' symptoms —fever, malaise, pain and coughing— vanished. Patients regained their appetites and grew more lively. But X rays revealed that while its symptoms had disappeared, the disease itself had spread. Researchers later discovered that cortisone had a similar impact on the course of other diseases caused by endotoxin-producing bacteria, among them typhoid fever.

Reflecting on the immune response to bacterial infection, the American scientist Lewis Thomas drew an analogy hauntingly appropriate to our times. "Our arsenals for fighting off bacteria are so powerful," he wrote, "that we are more in danger from them than from the invaders." Giving the military metaphor of disease a fresh twist, Thomas

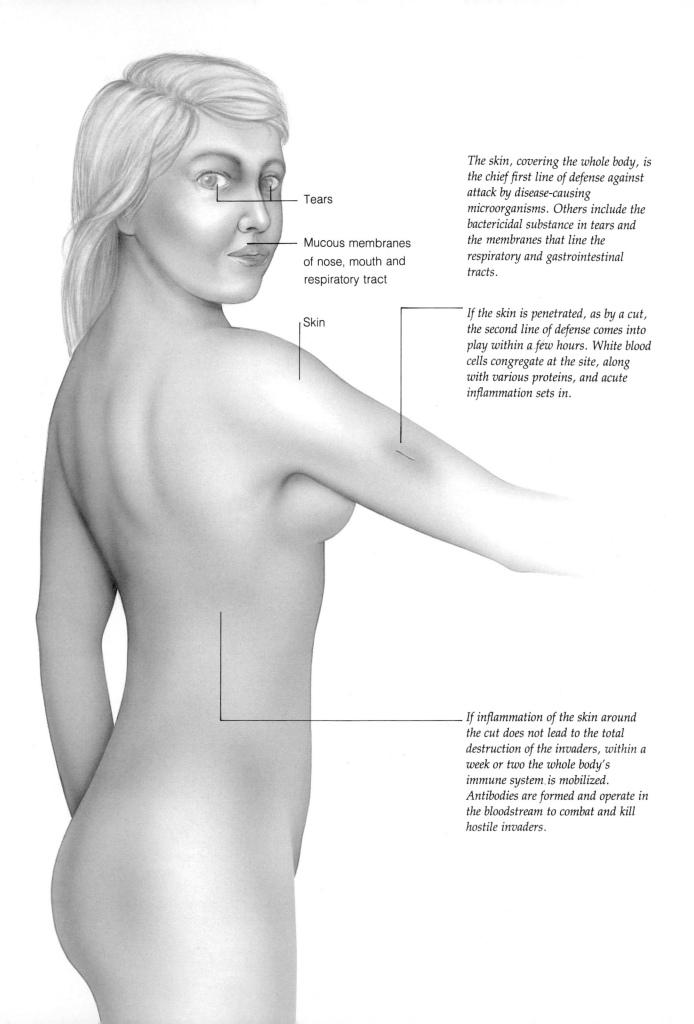

Tears

Mucous membranes
of nose, mouth and
respiratory tract

Skin

*The skin, covering the whole body, is
the chief first line of defense against
attack by disease-causing
microorganisms. Others include the
bactericidal substance in tears and
the membranes that line the
respiratory and gastrointestinal
tracts.*

*If the skin is penetrated, as by a cut,
the second line of defense comes into
play within a few hours. White blood
cells congregate at the site, along
with various proteins, and acute
inflammation sets in.*

*If inflammation of the skin around
the cut does not lead to the total
destruction of the invaders, within a
week or two the whole body's
immune system is mobilized.
Antibodies are formed and operate in
the bloodstream to combat and kill
hostile invaders.*

Paul Ehrlich

Discoverer of Magic Bullets

At the opening in 1906 of a research institute for chemotherapy in Berlin, a chemist prophesied a time when substances would exist that "would represent, so to speak, magic bullets" which would enable "complete sterilization of a highly infected host at one blow." The chemist was Paul Ehrlich, the inventor of chemotherapy.

He was born in 1854, in a country town in Prussian Silesia (now Poland). An unremarkable pupil, it was not until he went to university that he impressed his tutors, working late into the night on tissue preparations, modifying the dyes with which he stained them. His later development of staining techniques cut new paths to the identification of bacterial cells, and allowed him to discover the uses of the stain methylene blue, particularly to identify organisms that cause nervous disorders.

As Head Physician of the Charité Hospital in Berlin, he worked on the tuberculosis bacillus discovered by his friend Robert Koch. While studying it he contracted a mild case of tuberculosis which forced him to leave the damp German climate and go to Egypt for two years. On his return he received Koch's newly developed tuberculin

treatment which was extremely successful.

Puzzling over the problem of immunity, Ehrlich proposed a theory in which cells have receptors, or side chains, to which a toxin molecule anchors itself. If the cell survives, its receptors regenerate prolifically and prevent further infection, thus immunizing the organism. With Emil Behring, Ehrlich worked on a cure for diphtheria based on this theory, using antitoxins taken from animals which had been immunized against diphtheria. In 1892 they tried the serum on 220 sick children, with resounding success.

Ehrlich spent the rest of his life synthesizing his "magic bullets" from compounds of arsenic. He reached number 606 of the series, which he unsuccessfully tried to use against trypanosomes (the organisms that cause sleeping sickness). He was about to call a halt to the experiments but his Japanese colleague Sahachiro Hata showed him that 606 had outstanding results when used against syphilis in laboratory animals. In April 1910 Ehrlich announced the cure for syphilis. His reluctance to release the compound for mass production because he was still unsure about its side-effects could not quell the clamor it provoked. By the end of the year 65,000 units of it were manufactured, under the name Salvarsan (arsphenamine). His triumph was marred by skeptical accusations of charlatanism.

The strain of Ehrlich's success caused his already frail health to worsen. In 1914 he had a stroke and, while recovering in a sanatorium a year later, had his second, fatal stroke. His stature in medical history was succinctly described in Sir Robert Muir's words, "Ehrlich must be with the greatest, however small that company may be."

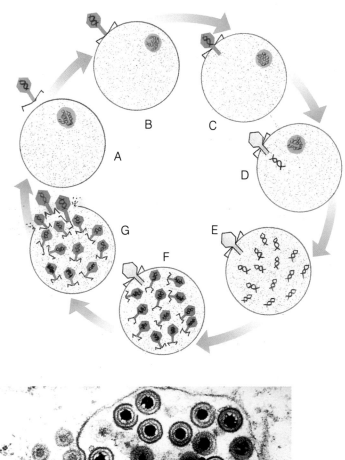

Viruses that infect bacteria, called bacteriophages, replicate inside a bacterial cell. The phage approaches the bacterium (A) and attaches itself to the cell wall (B). It inserts its tail through the cell membrane (C) and injects its genetic material (D). This material then replicates (E) and synthesizes new protein coats around each new section (F). In most infections the cell wall disintegrates, releasing new phages (G).

concluded that the body politic is "at the mercy of our own Pentagons, most of the time." On the other hand, the body nearly always wins.

As well as toxins, bacteria produce enzymes and other substances which enhance their virulence. The enzyme hyaluronidase is produced by several bacteria called cocci, in response to the presence of hyaluronic acid, one of the compounds of ground substance (the "mortar" which binds cells to build tissues). By dissolving hyaluronic acid, hyaluronidase hastens the spread of bacteria in host tissues. *Clostridium perfringens*, a resident of the intestinal tract, produces phospholipase, an enzyme that dissolves cell membranes and destroys both tissue and red blood cells (it is also found in the venom of some poisonous snakes).

Some staphylococci, members of the "normal flora" which frequently turn nasty to cause boils, carbuncles, abscesses and septicemias, produce coagulase. This enzyme thickens fibrinogen, the protein that weaves strands of fibrin to make blood clots. The staphylococci wrap themselves in fibrin as a camouflage against killer cells of the immune system. Other staphylococci and some streptococci protect themselves with leucocidin, a substance that kills leucocytes (white blood cells). Several sorts of bacteria produce hemolysins, substances that enhance the virulence of the bacteria by enabling them to destroy red blood cells. Some pneumococci are enclosed within capsular material which, although not toxic, protects them against the predations of leucocytes. Thus there are many strategies that enable bacteria to invade and spread within host tissue as well as to elude and resist attacks from the immune system.

Magic Bullets

From early times man has sought to blunt the weapons and foil the stratagems of bacteria with chemical agents. But too often chemical treatment damaged or destroyed healthy cells and tissues and disrupted or weakened the body's defenses in the process of killing the microbial parasites. To serve as an effective therapeutic agent, a chemical must

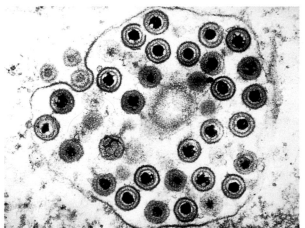

The Herpes simplex *virus belongs to a family of viruses which give rise to many diseases, including genital and oral lesions, encephalitis, shingles and chickenpox. The name herpes was once a generic term for all viral infections that induced erupting vesicles on the skin.*

Viral shapes range widely, from spherical to filamentous. Some, such as influenza, even have various forms, from spherical to ovoid. The bacteriophages (T) have a hexagonal head and a tail structure.

The virus HTLV III, identified as the likely cause of the disease AIDS (acquired immune deficiency syndrome) transmitted via blood, saliva or semen, is seen in the electron micrograph below.

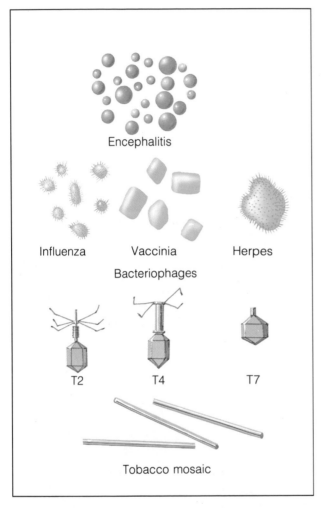

Encephalitis

Influenza   Vaccinia   Herpes

Bacteriophages

T2   T4   T7

Tobacco mosaic

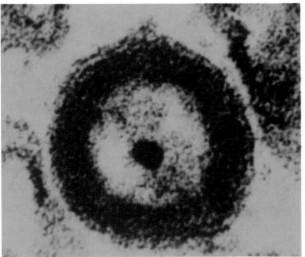

infiltrate the cells and tissues of the host without harming them, seeking out and destroying or crippling the offending microbes. Such a chemical is known as a "magic bullet."

Two kinds of "magic bullets" have been cast, one compounded from chemicals and the other drawn from microorganisms. About 1910, Paul Ehrlich, a close friend and colleague of Robert Koch, developed Salvarsan (arsphenamine), an arsenic compound which proved an effective treatment for syphilis. But further progress in chemotherapy was not made until 1935 when another German, Gerhard Domagk, showed that a group of compounds known as sulfonamides worked well against a variety of bacteria although they could not be targeted accurately against specific microbes. Within a decade of Domagk's discovery, spurred by the exigencies of war, scientists had developed some 5,500 sulfonamides and tested them for antibacterial activity. These "sulfa drugs," as they are called, are still used to treat a wide range of bacterial infections, although they have been superseded by the second group of chemotherapeutic agents—antibiotics.

Antibiotics were originally derived from living organisms, and their effectiveness depends on the fact that certain organisms — such as some types of molds and bacteria — produce chemical substances that kill or inhibit the growth of disease-causing bacteria. This antagonism and its effects found application in medicine long before they were proved in clinical trials. Centuries ago the Chinese were using moldy soybean curd to treat boils. But scientists began searching systematically for antibiotics only during the 1920s.

In 1928 Sir Alexander Fleming, an English bacteriologist, found by chance that one of his colonies of staphylococci had become contaminated by a mold around which no bacteria grew. It took many scientists working in England and America more than a decade to isolate the specific antibacterial element from the mold and turn it into a safe and effective drug. As Fleming identified the mold as *Penicillium* sp., the new "miracle drug" was dubbed penicillin; it is the original and still the most generally useful antibiotic. Since Fleming's original discovery, penicillin G, many closely related penicillins have been developed including synthetic ones, some of which destroy not only those bacteria sensitive to penicillin G but other types of bacteria as well.

Since 1940 several thousand antibiotic substances have been isolated, many of them from microorganisms living in the soil. Streptomycin, discovered in 1944, destroys many microbes resistant to sulfonamides and penicillin. Erythromycin, which was discovered in 1952, is especially useful because it affects microbes that have become resistant to penicillin and streptomycin. Because different antibiotics attack different microbes, they are often used in tandem.

Resistance to chemotherapeutic agents not only hampers the treatment of disease but also encourages the incidence of new ones. Prolonged use of antibiotics may lead to the replacement of "normal flora" by microbes resistant to whatever drug the patient is taking. These resistant microbes then cause a secondary infection, or superinfection, alongside the original illness. Other drugs, especially those intended to limit the immune response, may cause members of the "normal flora" to turn pathogenic themselves. Other factors — nutritional deficiencies, shock, trauma, and the presence of other diseases — may increase the likelihood of infection by previously benign, resident microbes. Ironically, the progress of medical science has eliminated from most parts of the world the danger from many of the dreaded diseases of the past — plague, cholera, anthrax and the like — but has increased the threat from those microbes which have long found their home in man.

The Study of Viruses

Viruses, the second major agents of infectious disease, are responsible for many human diseases, including the common cold, influenza, chickenpox, smallpox, poliomyelitis, hepatitis, herpes and yellow fever. Some so-called "slow viruses" may have an incubation period of months, even years, between initial infection and the appearance of symptoms. Among these are the relatively uncommon, but fatal Creutzfeldt-Jakob disease and kuru, a degenerative disease found only among a few tribes in a remote area of central New Guinea.

Until the 1930s, viruses were studied indirectly through the diseases they caused. Serum drawn

Working in his office at St Mary's Hospital in London, in 1928, Sir Alexander Fleming accidentally stumbled on penicillin. He had carelessly left a dish of staphylococci uncovered and noticed later that the culture contained empty patches. Studying them more closely he found that a mold, Penicillium notatum, had gotten into the dish and destroyed the bacteria in certain areas. The mold was later shown to produce penicillin, the first antibiotic.

from patients who had recovered from viral infections contained antibodies formed in response to specific viruses. These could be used to identify the viruses themselves. In 1935, Dr Wendell Stanley of the Rockefeller Institute for Medical Research, working with the highly stable tobacco mosaic virus, isolated a pure virus for the first time. By showing that the virus could be crystallized and that the crystals could produce disease, Stanley kindled the longstanding controversy over whether viruses were genuine living organisms. This controversy turns more on definitions of life than on characteristics of viruses.

Viruses are obligate parasites. Lacking much of the machinery that allows other cells to take in nutrients and use them to grow and multiply, they can reproduce only by infecting the living cells of host tissue. Viruses appear to be alive, unlike other chemical substances, because they can replicate themselves. The question of whether or not they actually are alive has never been settled to the satisfaction of all scientists and has been abandoned by science in favor of more fruitful research. Perhaps André Lwoff of the Pasteur Institute had the last word when he remarked in 1957 that "Viruses should be considered as viruses because viruses are viruses." The uniqueness of these organisms makes them, in Stanley's words "a scientific Rosetta stone" on which is etched the key to the relationship between inert molecules and living beings.

Viruses are more pebbles than stones, for they are very, very small. They range in size from about 25 to 400 nanometers in diameter (a nanometer, nm, is one billionth of a meter). It would take about 150,000,000 polio-causing viruses (each of which is about 25nm across) to cover the period at the end of this sentence. Bacteriophages, those viruses which infect bacteria, are approximately the same size as the smaller types of normal viruses, having diameters of between 25 and 100nm.

Thanks to their minute size, viruses escaped the microscopes and filters of early bacteriologists, coming into view only under the electron microscope in 1939. Their structure is simple. All viruses consist of nucleic acid, either deoxyribonucleic acid (DNA) or ribonucleic acid (RNA) but, unlike all other cellular forms of life, never

both. The nucleic acid is surrounded by a coat (capsid) made of proteinacious subunits (capsomeres). Some viruses also have internal proteins, and an outermost envelope made of lipoprotein. The protein in the capsid, not the nucleic acid, characterizes the identity of a virus. A virion (a complete virus particle) may contain one of four types of nucleic acid, single or double-stranded DNA or RNA. Viruses most likely to cause disease in man have been found to carry all types except single-stranded DNA.

Viruses have many shapes. Some are regular and many-sided (polyhedrons) with 20 triangular faces and twelve corners. Others are helical, or shaped like long rods or cylinders. Some polyhedral or helical viruses are wrapped in a flexible envelope, which may display varying shapes. Most appear spherical, like the herpes simplex virus.

Like bacteria, only a few viruses threaten human health. Scientists believe viruses may be genetic engineers, transferring genes between one organism and another. A striking example of the ability of viruses to cause change in living things is the tulip break virus's power to streak flowers with stunning hues. In seventeenth-century Holland, the value of "diseased" bulbs touched off a frenzy of speculation, bringing financial chaos that only political intervention could resolve. The American scientist Lewis Thomas likens viruses to hostesses at a party, carrying trays of genetic cocktails and hors d'oeuvres, occasionally spilling or dropping them to cause disease.

Viruses are unable to flourish and procreate by themselves so they commandeer the resources of cells, exploiting their enzymes to generate energy, manufacture proteins and produce more viruses. Their nucleic acid carries genetic instructions for producing more viruses and serves as a kind of requisition order to cells. Viral replication often destroys the host cell. Which cells are destroyed, and how many, determines whether or not a viral infection causes disease.

Replication begins when a virus attaches itself to the cell membrane, often using protein receptors that are studded over the outer surface of most viruses. In some cases the cell latches on to the virus just as the virus attaches to it.

Viruses breach cells by different routes. At one

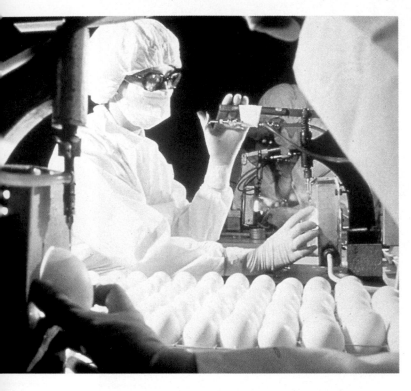

time, scientists supposed that cells swallowed or engulfed viruses by phagocytosis, an important technique among cells of the immune system. It has now been found that enveloped viruses fuse their lipoprotein envelope with the cell membrane, dodge defensive enzymes released by the cell and empty their nucleic acid into the cell's cytoplasm. Naked viruses — those without envelopes — do, however, seem to enter the cell by endocytotic pathways — the cell apparently engulfs the virus particles by surrounding them with part of the cell membrane and transporting them into the cell.

Inside the cell, viruses lose their protein coats, probably with help from cellular enzymes, and begin the processes that synthesize viral proteins and replicate viral nucleic acids. Viral proteins are required to build the new viruses being produced and to serve as enzymes for metabolizing nucleic acids. Once the protein coats and nucleic acids are synthesized in different parts of the cell, they are assembled into virus particles. Enveloped viruses acquire their envelopes from the cell's nuclear membrane as they make their way out of the cell — like picking them up from the cloakroom.

Different viruses leave the cell by different routes. Sometimes the invasion, occupation and exploitation of the cell ends with a bang as the cell bursts, releasing the viral progeny. Other viruses replicate without destroying the cell.

Viral infections are easier to prevent than to treat. Vaccines have proved to be the surest safeguards against viral disease. Neither Jenner, who established their use, nor Pasteur, who named them, understood how vaccines work. The small-pox and rabies vaccines they made contained "live" viruses, weaker than the viruses that caused the diseases. Pasteur produced his vaccine from "attenuated" rabies virus particles, weakened by generations of breeding. Although weakened, the viruses in vaccines are still sufficiently like their virulent counterparts and forebears to goad the immune system into producing antibodies.

Early efforts at producing vaccines foundered because viruses, by their very nature, are impossible to grow outside living cells. Pasteur cultivated his viruses in living animals, a slow and unreliable process. Just before World War I, tissue culturing was developed, which enabled pieces of tissue to be grown outside living animals; the live tissue could then be infected with viruses. This technique produced viruses in sufficient numbers to manufacture vaccines, but the cultures were not easily protected against contamination by bacteria. Only with antibiotics did this procedure yield more useful results.

Pure uncontaminated cultures were first obtained by growing viruses in chick embryos. Eggs incubated for a week or two can be inoculated with a virus injected through the shell. After the virus is introduced, the shell is sealed and the egg incubated until the virus matures. Smallpox, yellow fever, some influenza and other vaccines are still produced by this method, although tissue culturing is the most effective means for growing other viruses for commercial vaccine production.

Elusiveness — the Viral Nature

Apart from problems associated with culturing viruses, scientists searching for vaccines have been bedeviled by the genetic versatility of some pathogenic viruses. As winter nears, doctors and scientists nervously await the coming of influenza,

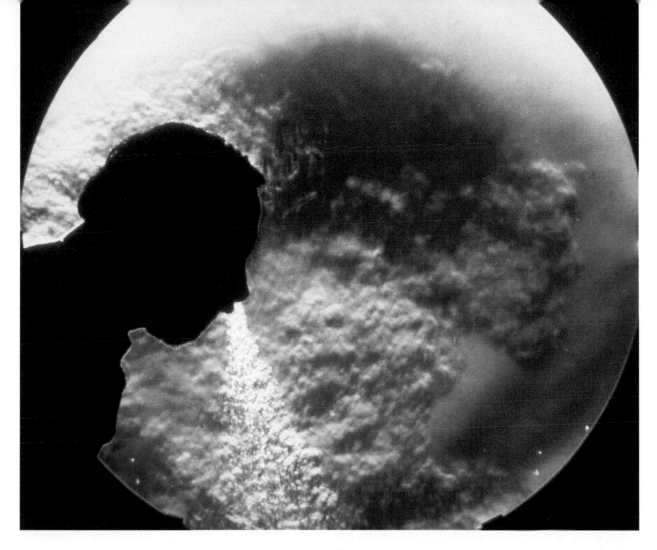

wondering what guise it will take. Influenza viruses fall into three major types — A, B and C. Types A and B, but not apparently type C, mutate — type A every two or three years and type B every three to six years — changing enough to appear to the immune system as a new virus, against which it is defenseless. The mutation of viruses is called antigenic drift, or when a significant change occurs, antigenic shift. Scientists think antigenic drift and shift arise from cross-breeding, or recombination, between different strains of human, mammal or bird viruses.

Only type A influenza viruses seem to undergo antigenic shifts profound enough to cause world-wide epidemics (pandemics). The Spanish flu of 1918, Asian flu of 1957 and Hong Kong flu of 1969 all originated from antigenic shifts in type A viruses. The pandemic of 1918 took 20,000,000 lives, higher than the toll of the war it followed. Until recently, virologists and epidemiologists enjoyed some success at predicting the influenza epidemics which struck roughly every three years. By analyzing cultures from sufferers during the winter and collecting data from the Southern

Hemisphere, where the flu season falls during the Northern Hemisphere's summer, they could prepare effective vaccines.

Lately, however, influenza viruses have begun to behave differently. Instead of widespread out-breaks occurring every three or four years, the disease spreads less widely but strikes more often. Mutant viruses no longer sweep around the world but spring up randomly in "asynchronic" epidem-ics. A strain may strike Europe but only gently nudge the United States. This sudden randomness of the influenza viruses puzzles scientists, although one possible explanation is increased airplane travel, which would have the effect of spreading new strains of disease rapidly and randomly.

Vaccines administered in more tranquil times can still prove effective. Researchers at the Center for Disease Control in Atlanta recommend that the elderly, the young and those suffering from chronic disease are given an initial vaccination followed by annual boosters containing the strain expected to appear. If the expected strain does strike, influenza vaccines offer protection to some 75 percent of persons who have been vaccinated.

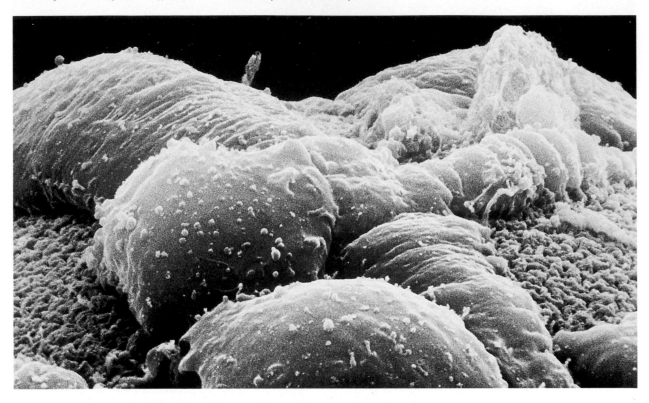

Chemical treatment of viral disease was long thought impossible, since no known therapeutic agent could distinguish between the living cell and the virus and so destroyed them both. Agents that kill viruses under laboratory conditions are easy enough to find but, when introduced into the body, they damage the cells in which the viruses replicate. Moreover, viral infections set in quickly, peaking just as their symptoms appear. It is therefore difficult to make accurate diagnoses of particular infections in time to prescribe appropriate medication. In addition, different viral infections display similar symptoms, further hindering accurate identification of the infection and timely diagnosis.

Despite the obscurity of the ways in which they operate, several agents are now being used to treat viral diseases. Acyclovir has proved effective in treating herpes; it shortens the time for which symptoms are present and lengthens the period between the recurrent outbreaks of symptoms. It does not cure the disease, however. Amantadine both prevents and controls certain influenzas, though the rapid spread of the disease often renders the drug useless. Methisazone prevents smallpox, but the disease may have been eradicated just as the drug was discovered. And interferon, apparently ineffective when used alone, may yet be used as an antiviral agent if it can be used in combination with another agent. This substance (discovered in 1956 by the British virologist Alick Isaacs) is a protein which is naturally produced by the body to act against diseases, such as viruses. Even though there has been some success in treating viruses, the causes of some diseases such as certain cancers, which may be viral, have yet to be established before researchers can begin to look for cures.

In a mere century, since Pasteur and Koch confirmed the germ theory of disease, science has driven myth and magic from the diagnosis and treatment of many diseases. One disease after another has been found to arise from a single, determinate source, often a bacterium or a virus, confirming Pasteur's remark that "The role of the infinitely small in nature is infinitely great."

Alick Isaacs

Interferon or Misinterpreton?

In 1956 Alick Isaacs announced the discovery of interferon, an antiviral substance produced naturally by the body. At that time there was no direct treatment for viral infections and so the revelation of this substance was enthusiastically welcomed as a tremendous breakthrough. Ever creative at such times, the skeptics termed it "misinterpreton."

One of four sons of a poor Glaswegian family of Lithuanian origin, Isaacs had, from a very young age, always expressed the desire to be a doctor and performed his first chemical experiment with a toy chemistry set. He excelled at school and won many prizes at university. He graduated with a medical degree but soon found that he was not interested in clinical medicine, and turned to bacteriology. When he accepted a fellowship to do research on influenza, he took a step that was to determine the path of his work for the rest of his life.

Isaacs was sent to Australia to study influenza and, in particular, an epidemic that had broken out on Ocean Island. The outbreak had apparently been introduced by Chinese workers from Hong Kong. While there he became interested in the response of the human body to the virus.

When he returned to England, Isaacs worked in the laboratories of the World Influenza Centre where he found evidence that the influenza epidemics that struck Britain in 1951 consisted of two strains: one originated in Scandinavia and had lain dormant from the previous winter; the other had come from the Southern Hemisphere.

Isaacs was fascinated by an interference phenomenon that had been known about for years— a cell infected by one virus cannot be infected by a second virus. Working on this phenomenon with the Swiss virologist Jean Lindenmann he found that a special protein was produced by animal cells when they were attacked by viruses—they called it interferon. Isaacs studied this substance for the rest of his life, trying to identify its chemical and physical properties and poring over the problems of its production.

Interferon is normally produced in tiny quantities and is extremely difficult to purify. As a result it was not possible to test its therapeutic potential until the revolution in science brought about by the advent of molecular biology. Large quantities can now be produced relatively cheaply and in sufficient quantities to study its possible use to treat cancer. In addition, interferon may be useful in preventing colds and influenza.

After a series of bouts of depression and mental disturbances, Isaacs developed a hemorrhage which eventually killed him in 1967. Dr Sam Baron gave him a worthy tribute which is a philosophy any scientist would do well to adopt, when he said of Isaacs that he "had the necessary faith in his conclusions to defend the most controversial ones until they were accepted or until new information convinced him that the proposed explanation was wrong."

Chapter 6

The Cell in Evolution

Life on Earth stretches back more than three billion years. From the moment life began, an unbroken chain of cell life-cycles has continued to connect every living thing in one vast family tree. Successive cell divisions and cell fusions link us to the first primitive cell, our ultimate ancestor. Thus indirectly, we are linked with dinosaurs, tapeworms, trees, mosses, mushrooms, seaweeds, and even bacteria. These are, in fact, our cousins, many, many times removed.

Evidence for Evolution

The evidence for evolution is preserved not only in fossils but also in the structures of living cells and the molecules from which they are made. Today, biologists can work out the exact arrangement of molecular building blocks in genes, which determine inherited characteristics. By comparing related genes in different organisms, they can build up "family trees" for individual genes, including ones that control those numerous characters not preserved as fossils.

A gene is that section of a DNA (deoxyribonucleic acid) molecule which codes for the structure of a specific protein molecule. DNA is a chain of nucleotides, and proteins are strings of amino acids. The order, or sequence, of the nucleotides determines (by the universal rules of "the genetic code") the sequence of amino acids in the proteins. By knowing these rules and the nucleotide sequence of a given protein's gene, scientists can determine the amino acid sequence of the protein. The nucleotide sequence can be deduced by extracting the DNA, then purifying specific genes from it by cloning — that is, by inserting the gene into pieces of bacterial DNA and allowing it to replicate inside living bacteria.

Such methods have a double value. By finding similarities between related genes, family relationships between different organisms can be traced in detail. And by seeking differences between related

The biblical vision of the end of the Earth with fire and brimstone hailing down from the heavens and the ground being rent asunder is, ironically, as conditions are imagined to have been during the beginning of life on Earth. It is thought that lightning bolts discharging a high voltage into the gases and water vapor of the early atmosphere created the essential chemical components of life and made the development of cells possible.

117

genes, the mutations that have occurred in the genes can be identified precisely. Mutations are rare changes in the structure of DNA and are the underlying cause of all evolutionary change.

The new science of molecular biology is helping to settle many of the old controversies about evolution, such as the independent evolution of eyes in the different animal groups or phyla. It turns out that the purple protein rhodopsin, which enables your eyes to see these words, evolved long before animals — it is present even in unicellular algae, simple plants that use it to detect light so that they know which way to swim.

The cloning of genes from living organisms is a powerful biochemical technique, but is it possible to clone genes from fossils? In older fossils the DNA has been totally destroyed, but for a few recently extinct species this may not be so. Genes from an extinct animal were first successfully cloned in 1984 by Alan Wilson and his colleagues. They obtained DNA from museum specimens of the quagga, a relative of zebras and horses, which became extinct in 1883. Further studies along these lines will be fascinating, even though they may not reach sufficiently far back to clarify large-scale evolution.

Even bacteria, the simplest cells, usually have a few thousand different genes, but unicellular protozoa, which probably evolved from bacteria 1,500 million years ago, have even more — between 5,000 and 40,000. Multicellular animals, which evolved from protozoa, have from 5,000 to more than ten times that number.

From Single Cells to Multicellular Bodies

For the first 2,800 million years (80 percent of the history of evolution) life consisted of single procaryotic cells (which are characterized by the lack of a visible nucleus), or simple microscopic chains of such cells. The relatively sudden evolution of multicellular organisms, a mere 600 to 700 million years ago, was caused by mutations that made protozoan cells stick together, differentiate from each other (or become specialized) and cooperate in building a multicellular body. Soon afterward further mutations produced all the main animal body types including worms of various sorts, mollusks, jellyfish, crustaceans, sea urchins and eventually vertebrates.

118

There are obvious differences between certain cell types in the different animals. For example, the rod cells in the retina of vertebrates' eyes are structurally quite different from the sensory cells in insects' eyes. They must have arisen independently from structurally less distinctive cells by mutations of the same general sort that alter the cells of the unicellular protozoa. The different body plans of insects and vertebrates are not really caused by differences in particular cell types. Instead they arise during embryonic development through differences in the way the same basic types of cell become arranged into tissues and organs.

To make this clearer think of the differences between a skyscraper office block, a concert hall, and a school. All may be built of the same type of brick, steel girders, wood and glass, just as all animals are made from various basic types of cells — epithelial cells, blood cells, muscle cells, and so on. The differences between the buildings arise from the way the bricks and girders are fitted together to make pieces of differing size and shape. So it is with animal development; similar cells can be put together to make a worm, a fish, a bird, or a human.

But there is an important difference between the way buildings are made and the development of cells. The cells themselves are the workmen, creating their own materials and sticking to each other with the help of natural biological glues. They can also change their own shape, and some animal cells even migrate from one part of the body to another to create new structures. Neither is there a separate architect's plan. Instead the design lies in the cells' DNA (of which they all have a copy), the template that controls the structure of proteins that make up the cell.

Mutations Cause Biological Diversity

Understanding cell evolution is, indeed, the key to understanding evolution because all living things are made of cells. To understand either the sudden origin of the different phyla of animals or plants, or the much longer evolution of their single-celled ancestors during the preceding 2,800 million years, it is necessary to understand how mutations in genes can cause changes in the structure and in the various functions of cells.

119

The discovery of the superbly preserved fossils of Archaeopteryx, *an animal which was half-reptile, half-bird, allowed scientists to link the evolutionary development of birds from a branch of reptiles.*

Evolutionary change depends on mutations in DNA and on the perpetuation and multiplication of these mutations by DNA replication. But the effects of mutations vary greatly in magnitude. Mutations that drastically change body structure rarely survive, because usually they produce grossly defective creatures that cannot compete successfully with the existing types. Most viable mutations thus result from trivial changes in molecular structure, noticeable only to molecular biologists and biochemists.

Evolution involves long periods of relatively slight change, probably punctuated by much more drastic transformations when a major new body-plan arises. Soon after each major successful body-plan change, the descendants of the successful type multiply rapidly and diversify to produce numerous variations on the basic theme.

The best example is the sudden appearance of most metazoan animal phyla in the early Cambrian, 530 million years ago. None of them could have evolved before the first successful multicellular animal; but as soon as it did evolve, mutations would have produced an immense variety of different cell arrangements. The most successful variants would have survived as ancestors of new phyla. But once the main possibilities had been tried out, and all habitats available for multicellular animals successfully colonized, innovation would inevitably have slowed down. Further new types would find it much harder to survive and a long period of slow change would set in.

Studies of living organisms are more helpful than the fossil record in understanding the speed of change within major innovations. This is because big evolutionary steps tend to take place in small localized populations, in which intermediates are so rapidly replaced by their changed descendants that there are few of them to have any chance of becoming fossilized. Most of the members of these groups in any case were minute and had no hard parts that could readily fossilize. Only rarely does luck provide genuine missing links between major groups, such as *Archaeopteryx*, the half-bird, half-reptile. Most fossils come from large populations that persist for immense geological periods after the new types have spread widely around the world.

Charles Darwin (1809–1882), who first led most scientists to accept the theory of evolution, thought that fossils began abruptly with the first hard-bodied animals at the beginning of the Cambrian, 530 million years ago. But with the help of microscopes modern paleontologists have discovered fossil simple cells in rocks of all ages back to about 3,500 million years ago.

So the immense period from 3,500 million to 650 million years ago was not, as Darwin guessed, one of gradual evolution of multicellular creatures. It was simply the age of unicellular life. But how were the first cells formed? How, when, and where did life begin?

Vitalism versus Mechanism

There have been two schools of thought, both traceable to the ancient Greeks and to Asiatic religions and philosophies. According to one, mind is independent of, and comes before, matter — a universal mind or God created the universe and living things. The second, materialist, school of thought is that matter is fundamental and that minds came only later — living things and human minds evolved spontaneously from non-living and non-spiritual matter. Although many modern scientists are caught, in some ways, between both points of view, the materialist or mechanistic view

Charles Robert Darwin

The Origin of Species

One of the most outstanding naturalists, Charles Darwin is familiar throughout the world for his contribution to the understanding of evolution. He was born on February 9, 1809 in Shrewsbury, England, and was grandson of Erasmus Darwin the physician, philosopher and poet.

At school he was slow to learn; later he described himself at that time as "given to inventing deliberate falsehoods", which he always did "for the sake of causing excitement." Having abandoned medicine at Edinburgh University and having briefly considered a career in the Church, he took up natural history. He found the subject immediately fascinating, as is made evident by his recollection of University life that "no pursuit at Cambridge was followed with nearly so much eagerness or gave so much pleasure as collecting beetles."

On the recommendation of John Henslow, Professor of Botany at Cambridge, Darwin was given the position of naturalist aboard HMS *Beagle*, and set sail for South America and the Pacific in 1831. Initially concerned with the geological aspects of the countries they visited, he was struck by the diversity of animal species.

In the Galapagos Islands, he was intrigued by the finches. He recorded 14 different species, each one adapted to its own ecological niche. He later reasoned that they must have evolved from a parent species of finch from the mainland of South America.

After five years of traveling, Darwin returned to England and recorded his travels in *A Naturalist's Voyage on the Beagle*. He also published several books reflecting his interest in geology which established his name in scientific circles. He married his first cousin Emma Wedgwood and, despite continuous illness (now thought to be Chagas' disease), fathered ten children.

For more than 20 years Darwin amassed evidence to support his theory that the process of evolution rested on natural selection. He believed that the "death of a species is a consequence of non-adaptation to circumstances." He recognized that species are not fixed and are constantly undergoing modifications (or evolution), and that one species may evolve from another species.

Having reached these conclusions Darwin was stunned by a letter he received from Alfred Russel Wallace in 1858 outlining very similar theories on evolution to his own. A dispute over originality was solved by both men presenting a joint paper to the Linnean Society in that year. The publication of Darwin's *On The Origin of Species by Means of Natural Selection* in 1859 outraged orthodox scientists and churchmen, but Darwin left its defense to his close friend Thomas Huxley who declared at a famous Oxford debate in 1860 "*Genesis* is a lie, the whole framework of the Book of Life falls to pieces, and the revelation of God to man, as we Christians know it, is a delusion and a sham." Darwin probably would not have agreed, but went on to apply his theories to the evolution of humans in *The Descent of Man*, tracing our ancestors to the apes, which brought him fame and created much mirth among skeptics.

In later life Darwin turned to the plant world, to which he applied his theories of evolution with equal vigor. He was buried in London's Westminster Abbey on April 26, 1882.

The vitalistic image of the creation of man from a lump of clay, molded by a Creator, contrasts sharply with the pragmatic view that molecules and biochemical reactions determine the cellular formation of organisms.

Radioastronomy shows that molecules vital to the formation of life are forming spontaneously throughout the universe and accumulating, especially in the dense clouds of interstellar dust and gas.

has predominated in science from the Ancient Greek atomists up to the present day.

For centuries the vitalistic picture of life has been that of a creator molding clay into life-forms, like a potter, and then separately breathing life into them. This picture is contradicted by the modern molecular biological view about the fundamental nature of life. Life depends simply on the chemical and physical properties of the molecules of which cells are made. These material properties determine how the molecules react with and stick together to make functioning cell structures.

To explain how life began it is first necessary to explain how these particular organic molecules came into existence. Then it must be explained how, once formed, the nucleic acids, proteins and lipids, came together to form the first cell. The spontaneous origin of organic molecules is fairly well understood, and has a sound experimental basis. Explanations of the origin of life and the cell itself are still speculative, but are nonetheless at a much more advanced stage than they were 50 years ago, when the Soviet biochemist Aleksandr Oparin and the Englishman J. B. S. Haldane founded modern conceptions of chemical evolution.

The Origin of Organic Molecules

The vitalist argument that living organisms could not be formed only by organic molecules — that the hand of the Creator should be evident — lost ground rapidly after the synthesis of the organic compound urea from inorganic material in 1828 by the German chemist Friedrich Wöhler (1800–1882). For the next century, chemists concentrated on synthesizing more and more complex organic molecules in the test tube. From this long period of laboratory research, a basic chemical principle emerged to provide the fundamental clue to later ideas and experiments. This was the realization that the formation of organic molecules requires a certain amount of energy, but that energy can also cause the decomposition of organic substances.

In the laboratory chemists supplied the energy to inorganic molecules as heat or light. In nature plants get energy from light, animals from the chemical energy stored in food, which acts as a fuel as well as supplying the materials for making new cells. But supply too much energy, as in a forest fire or when a body is cremated, and in the presence of oxygen organic molecules decompose to carbon dioxide and water once more.

The stability of an organic molecule depends not just on the energy of its surroundings, but also on what other chemicals are present for it to react with. Oparin and Haldane independently realized that when an animal or plant dies, apart from any interaction with bacteria or even fungi it decomposes relatively rapidly simply because the abundant oxygen in the air makes the organic molecules unstable. Without oxygen, decomposition is far slower.

Oparin and Haldane suggested that if the early atmosphere contained little or no oxygen, spontaneously-formed organic molecules would therefore have decomposed much more slowly than they do today. They would have accumulated in the sea, lakes, ponds and soil to form a prebiotic soup, rich in organic molecules.

Nowadays a vast array of microorganisms — bacteria, protozoa and fungi — feed on organic molecules; no sooner are such molecules formed than something gobbles them up, decomposing them to carbon dioxide and water. Darwin himself pointed out that before life began this was obviously not the case, and the molecules would have slowly accumulated and provided the raw materials and food supply for early life.

How Life Created the Oxygen Atmosphere

Ancient sedimentary rocks from 3,000 million to about 2,000 million years ago contain several minerals, such as iron and uranium, which are only partly combined with oxygen. In today's atmosphere, these minerals would be rapidly oxidized to different forms by atmospheric oxygen. From about 2,000 million years ago, reduced minerals are no longer found in the rocks in such quantities but are replaced by highly oxidized ones, such as the red iron oxides that redden so many desert rocks. These vast "red beds" of ancient rock testify to a newly oxidizing atmosphere. To begin with, the Earth's atmospheric conditions alternated repeatedly between oxidized and reduced, producing "banded iron formations" — rocks with regular dark and light stripes signifying reducing and oxidizing conditions.

Today, most oxygen is produced by photosynthesis, the process by which plants trap light energy to manufacture food, and oxygen is left as a waste product. The evolution of photosynthesis is probably what changed the Earth's atmosphere overall from reducing to oxidizing 2,000 million years ago.

The first oxygen came not from plants — they evolved very much later — but from blue-green bacteria, the cyanobacteria, which are often (incorrectly) called blue-green algae. Unlike most bacteria, they are often multicellular organisms. A few are even visible to the naked eye and can be found in moist places; superficially they resemble seaweeds or algae.

Their cell structure is, however, different from that of true seaweeds or algae which, like plants and animals, all have cells with a nucleus, whereas the blue-greens do not. Because of these and other fundamental differences from typical nucleated cells, cyanobacteria are classified with other procaryotic cells, or procaryotes. Nucleated cells, which divide by mitosis, are called eucaryotic and the organisms they constitute — including animals, plants and fungi — the eucaryotes.

The difference between procaryotes and eucaryotes represents the most profound evolutionary divergence in the history of life. Unfortunately, cell nuclei are not preserved in fossil form, so it is often unclear whether a particular fossil cell was procaryotic or eucaryotic.

Usually the only clues are the size and shape of the cells and whether they are joined together in such a way as to make simple filaments or multicellular bodies. Procaryotic bacterial cells cannot grow to a large size, and so any really large

Plants trap the energy from sunlight and use it to make organic materials. This process, which gives off oxygen as a waste product, is the source of most of the oxygen on Earth. Oxygen is thought to be virtually absent from the other planets of the Solar System.

Boiling, putrid volcanic pools are seemingly uninhabitable, but some heat-resistant bacteria which do not need oxygen live in them, as do some that survive on the sulfur emanating from the pool.

fossil cells are probably eucaryotic. Such large cells first appeared about 1,500 million years ago, before which the most advanced forms of life consisted of procaryotic bacteria and there were no eucaryotic organisms.

The First Life-forms

Because free oxygen was severely limited during early evolution, the first bacteria must have been anaerobic — able to live without oxygen. Hundreds of different kinds of anaerobic bacteria now thrive in places such as stinking black mud, or the guts of animals, which contain little or no oxygen. They obtain energy and food in many different ways.

Some anaerobic bacteria simply feed on carbon dioxide and hydrogen and produce methane as a waste product. These bacteria are classified in the group Archaebacteria together with some odd types of bacteria that inhabit the most unlikely environments: salty lakes or seas (the Dead Sea, for example), hot acid waters such as those that ooze from heaps of burning coal, and the hot, alkaline waters of desert lakes. The fundamental cell structure of archaebacteria differs greatly from that of all other bacteria, which are called eubacteria. Although some scientists consider the archaebacteria to be the most ancient forms of life, it is probable that they are contemporary, specialized adaptations to odd habitats, rather than relics of earliest evolution.

Eubacteria include many anaerobes, the oxygen-producing cyanobacteria, and a vast array of non-photosynthetic aerobic bacteria that use oxygen to respire much as we do. Aerobic bacteria (those that use oxygen) could not have evolved until after cyanobacteria filled the atmosphere with oxygen. This gas is a highly reactive chemical, and is toxic to any cells that have not evolved protection against it, and this includes the anaerobes. Oxygen can be considered to be the Earth's first large-scale air-pollutant. Initially, respiration may have evolved as much to protect cells from oxygen poisoning as to provide useful energy.

Most aerobic bacteria feed on dissolved organic materials, just as fungi do. A few, however, obtain energy from inorganic sources, such as hydrogen, or reduced sulfur, iron or nitrogen compounds. They use this energy to convert carbon dioxide into organic molecules to make body tissues. When these chemosynthetic bacteria were first discovered last century some thought them to be the most primitive forms of life, because they could subsist entirely on inorganic food.

Haldane and Oparin pointed out, however, that chemosynthesis, respiration and photosynthesis are complicated processes, requiring many enzymes, which the bacteria did not have. They thought the first cells must have been simpler anaerobic bacteria that fed directly on the ready-made organic molecules of the prebiotic soup. In this they would not even need enzymes to synthesize the basic building blocks of life — amino acids, nucleotides and lipids — but could simply absorb them from the environment by non-biological mechanisms.

Many anaerobic eubacteria obtain energy directly from their food by a sort of fermentation process, much as a yeast fungus does when fermenting sugar to wine. These non-photosynthetic anaerobic bacteria were what Oparin and Haldane believed to be the most primitive cells. Recently, several scientists have questioned this idea and suggest instead that they may have been photosynthetic because some anaerobic eubacteria get their energy by photosynthesis. These are the green and the purple bacteria which, unlike plants and cyanobac-

Much of the nonvolcanic sulfur mined from rocks — as depicted in Renato Guttuso's **Sulfur Miners** *— was made by photosynthetic, or sulfur, bacteria. The first photosynthetic organisms, they produced sulfur as a waste. Such deposits occur in Sicily and around the Gulf of Mexico in the United States.*

Viruses are some of the simplest structures capable of self-replication. They have a protein coat, not a cell membrane, and the DNA is a loose strand floating freely inside rather than bound in a nucleus, as in eucaryotic cells. One form of virus, the bacteriophage (shown here), needs bacterial cells as a host in which it can replicate. Such viruses have a hexagonal head on a tail sheath with fibers, and hook onto a bacterial wall with the spikes on their base plates. They push the tail into the bacterial cell and release their DNA into the bacterium, where it multiplies and induces the bacterium to make new phages.

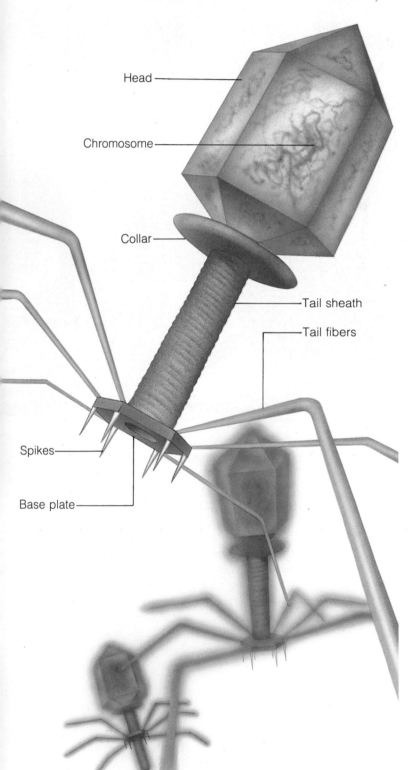

Head

Chromosome

Collar

Tail sheath

Tail fibers

Spikes

Base plate

teria, produce no oxygen but instead make use of hydrogen or hydrogen sulfide.

Many of the world's economically important yellow sulfur deposits were produced not by volcanoes but by green or purple bacteria, which are often called sulfur bacteria. They were the only photosynthesizers in the first 2,000 million years of life, and provided all the energy for the biosphere after the organic molecules of the prebiotic soup were consumed by early life.

The Origin of Life: Primitive Replication

Early living organisms are thought to have been more like viruses than true cells. A virus is a piece of DNA or RNA — the virus chromosome — encapsulated in a protein coat, which can infect cells and replicate inside them. But this presents biologists with a dilemma. The genes of the chromosome, made from nucleic acid, are necessary to make a protein; but a protein is necessary to make (that is, replicate) a gene. If a gene cannot be made without a protein or vice versa, how did they both come about? This is one of the most difficult remaining questions about the origin of life. How chemical evolution might lead to the formation of nucleic acids and proteins is well understood, but how genes came to code for and specify the structure of proteins is not, although there are plenty of ideas.

One possibility is that the first nucleic acids did not need proteins for replication. After all, proteins in the form of enzymes are only catalysts needed to speed up a reaction, and plenty of other chemicals can act as catalysts, such as metal ions. The idea that early life may have used zinc ions, found naturally dissolved in the oceans or on their rocky fringes, to catalyze its replication is supported by the fact that many modern nucleic acid enzyme proteins contain zinc.

Given even inefficient replication of a double helix of DNA, there would be a primitive form of inheritance, because the spontaneous pairing of the bases of nucleic acids would ensure that daughter double helices had the same nucleotide sequence as their parents. Reproduction and inheritance — the fundamental features of life (and evolution) — would have begun. Mistakes in base-pairing, which for chemical reasons inevitably sometimes

Harold Urey

Chemistry and Life

Harold Urey is perhaps best known for his discovery of deuterium, a heavy isotope of hydrogen. Among many other scientific advances, this discovery was instrumental in the development of the hydrogen bomb in the late 1940s. His theories on the origins of life and the formation of the Solar System are also of great importance. He is further renowned for his tables of elements and their abundances, compiled with Hans Suess. Substantial parts of these tables are still accepted today and, in fact, provide a basis from which physicists have studied the origins of the elements.

Urey was born in Walkerton, Indiana on April 29, 1893 and was the grandson of early pioneers. His father was a clergyman who died when Urey was six and he was brought up by his mother and his stepfather, also a man of the Church. He studied zoology at Montana State University but during World War I worked with high explosives, which introduced him to chemistry. Having obtained a PhD at the University of California, Urey set about studying the properties of heavy hydrogen.

The discovery of deuterium had a far-reaching effect on many fields of science. It made possible the accurate

measurement of the atomic weights of hydrogen and oxygen, and led to the discovery of oxygen isotopes. For biochemists it provided an alternative to ordinary hydrogen in compounds, and Rudolf Schoenheimer employed it as an isotope tracer in determining the chemical reactions in live tissue.

In 1934 Urey was awarded the Nobel Prize in chemistry "for his discovery of heavy hydrogen" (but refused to travel to Sweden to pick up the prize because his wife was pregnant).

He successfully prepared high concentrations of other isotopes, such as nitrogen-15 and carbon-13. Separation of the rare isotope uranium-235 was instrumental in the development of the deadly atomic bombs during and just after World War II.

Perhaps realizing the danger of his discoveries led Urey to concentrate on geochemistry, an area of less destructive potential. By recognizing certain differences in chemical properties he formulated a method by which the temperatures of ancient seas (paleotemperatures) could be calculated. He rightly deduced that the heavy isotopes of oxygen are more highly concentrated in calcium carbonate shells than in the water from which the shells were deposited and was, therefore, able to date shells from as long ago as the Jurassic period.

From there he turned his attention to the beginnings of life itself and concluded that life must have originated during a time of reduced oxygen on Earth. He suggested that ultraviolet light provided the energy for the prebiotic synthesis of nucleotides and amino acids, because it was the most abundant suitable energy source on the primitive Earth. Urey also realized the importance of lightning and other electrical discharges — the second most abundant source of energy.

It was in his laboratories that Stanley Miller showed that the chemical compounds assembled into the molecules of living organisms can be made by inorganic processes.

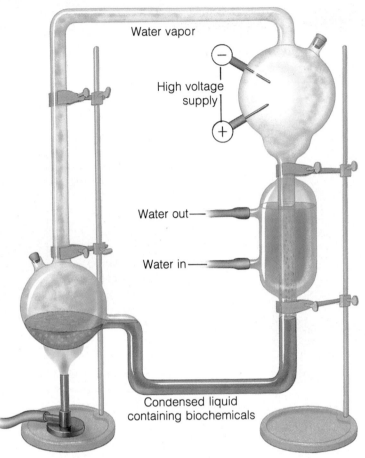

Stanley Miller tried to recreate the conditions on primitive Earth by sending electric sparks through water vapor and gases. The condensed liquid contained several of the substances essential for life.

Water vapor

High voltage supply

–

+

Water out

Water in

Condensed liquid containing biochemicals

occur, would produce occasional changes in the nucleotide sequence — the first mutations. At the next generation, these mutations would automatically be copied and could therefore be perpetuated generation after generation.

If new mutations made the nucleic acid more difficult to replicate, they would tend to die out. On the other hand, if they made replication easier, the new mutations would spread through the population of replicating molecules. Thus evolution and evolutionary progress would have begun.

Simulations of Evolution

The three key features of Darwinian evolution, inheritance, occasional mutations, and the resultant inherited differences in reproduction rate were all contained in the replicating nucleic acid molecules. Darwin referred to differences in reproduction rate as the "struggle for existence" and the overgrowth of some variants as "natural

selection," only in this case it was molecules, not living organisms, that were competing for survival and reproduction.

Many researchers, including Haldane, Oparin and the American chemist Harold Urey, proposed that ultraviolet light, which was the most abundant energy source on the early Earth, provided the energy for the creation of life. In 1953 Urey's student, Stanley Miller, performed a now classic experiment in which he sent an electrical discharge through four of the gases that would have been present in the early atmosphere and produced molecules of the essential components of life. Many other investigators have since done similar experiments and together they have yielded a great diversity of organic molecules: all 20 of the amino acids from which proteins are made; the bases and sugars from which nucleotides and nucleic acids are formed; fatty acids which become the lipids of cell membranes; and porphyrins that form the basis for chlorophyll, the substance needed for photosynthesis.

Scientists are therefore confident that all the necessary molecular building blocks would have been produced and were available to accumulate in the prebiotic soup. What is less clear is how they were combined to form nucleic acids and proteins. Some experiments have been conducted to answer this question. Researchers have found that amino acids and nucleotides can be combined (polymerized) in water, with the aid of special condensing agents. Many condensing agents are known, some readily formed under simulated prebiotic conditions.

The alternative solution is to heat the amino acids or nucleotides in the absence of water to polymerize them into proteins and nucleic acids. This could have happened when drops of prebiotic soup splashed on to tropical beaches where they then dried and roasted in the sun, or at the edge of volcanic lava flows where the heat was sufficient to promote the formation of proteins but insufficient to destroy them.

In experiments, tiny protein microspheres were created which, it was hoped, might represent some form of primeval unnucleated cell. Cell membranes, however, do not consist purely of protein, but of lipid and protein. Moreover, most biologists

Lightning, after ultraviolet light, was the most prolific source of energy on the early Earth. Electrical storms were probably far more frequent than they are today because of the abundance of clouds of water vapor in the air. Once the essential components of life had been formed and built up in the prebiotic soup, it is thought that great heat was needed to bind these components to make proteins.

The first single cell took about one billion years to appear, and a further two billion years passed before the first multicellular organisms evolved. Only in the last two million years has man evolved.

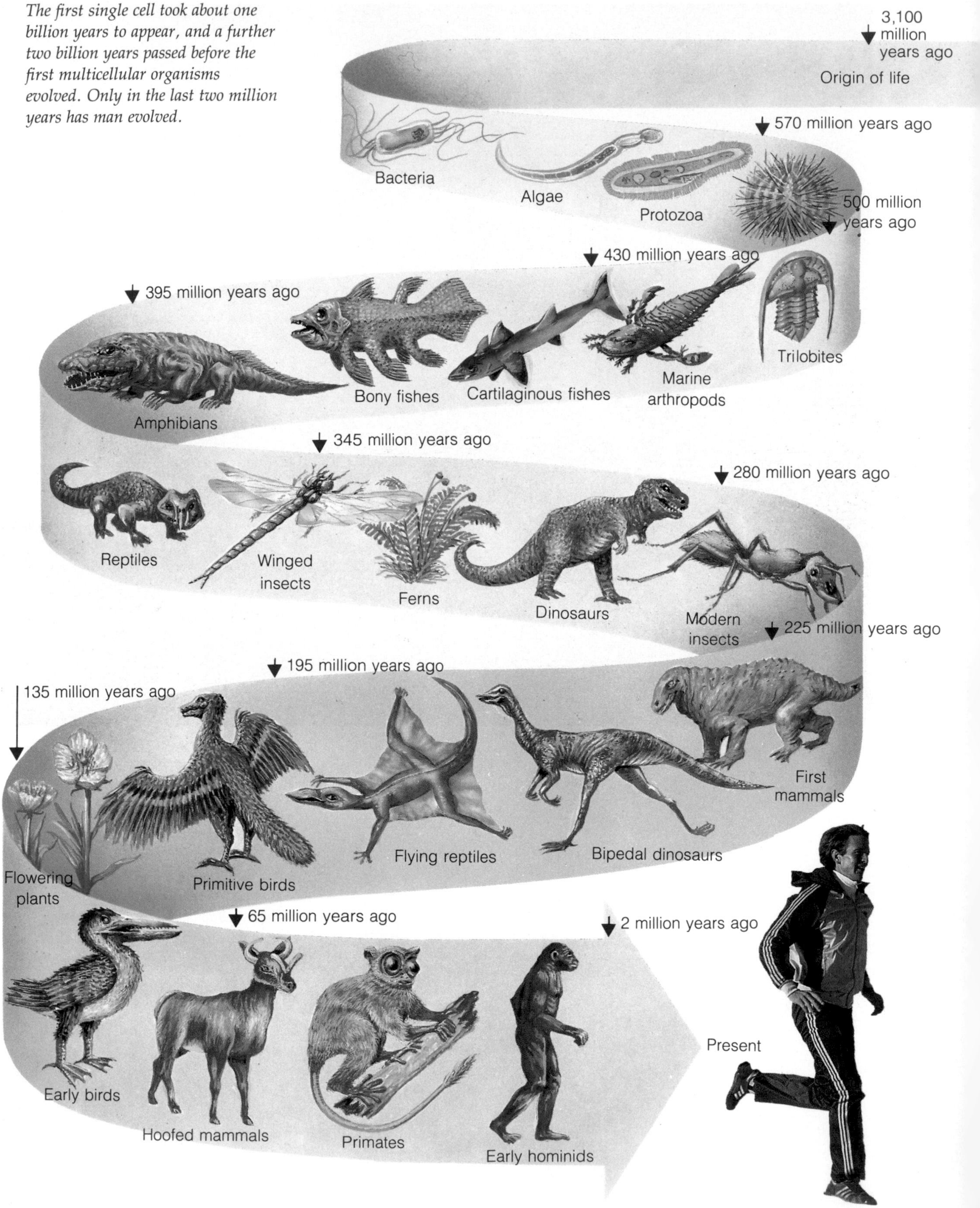

3,100 million years ago

Origin of life

570 million years ago

Bacteria

Algae

Protozoa

500 million years ago

430 million years ago

Trilobites

395 million years ago

Bony fishes

Cartilaginous fishes

Marine arthropods

Amphibians

345 million years ago

280 million years ago

Reptiles

Winged insects

Ferns

Dinosaurs

Modern insects

225 million years ago

195 million years ago

135 million years ago

First mammals

Flowering plants

Primitive birds

Flying reptiles

Bipedal dinosaurs

65 million years ago

2 million years ago

Early birds

Hoofed mammals

Primates

Early hominids

Present

132

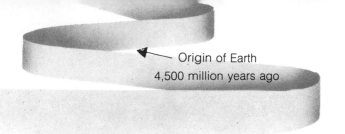

Origin of Earth
4,500 million years ago

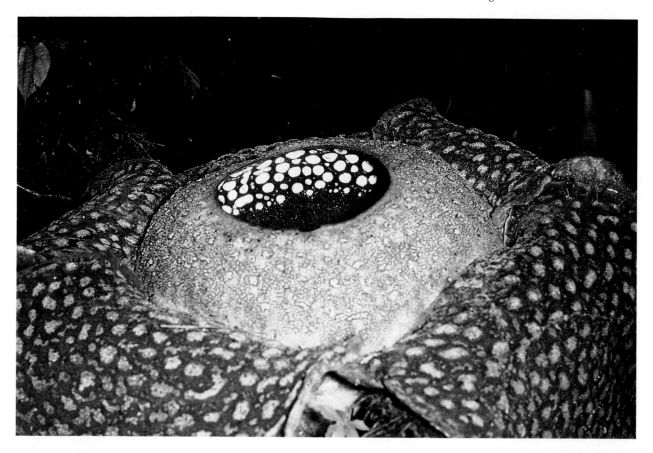

consider the lipid part to be structurally the more fundamental. The microspheres also have no nucleic acids, and it is hard to see how they could acquire them and evolve into cells. The jump from microspheres to cells seems too great. It is more likely that there was a period of replicating nucleic acids, unenclosed by membranes, before cells evolved.

The First Cell

The simplest cells must have had DNA, ribosomes (or their precursors), a plasma membrane, and the ability to grow and divide. They would have grown by absorbing molecules from the prebiotic soup outside and by making extra ribosomes and new lipid and protein molecules to expand their plasma membrane. Division would most likely have been by inward-furrowing of the plasma membrane, pinching the cell in two.

A plausible origin for the first cell was proposed in 1980 by Gunter Blobel of Rockefeller University. Blobel suggested that chromosomes and ribosomes first became associated with membranes by sticking to the outside of liposomelike vesicles. The simplest liposomes consist of a phospholipid two-layered structure just like the plasma membrane. The chromosomes and ribosomes would have had free access to all the amino acids and nucleotides in the external soup. This "inside-out" cell, as Blobel calls it, could then evolve sophisticated membrane proteins to help it grow and divide, transport nutrients, or provide energy for itself long before the chromosomes and ribosomes became trapped inside a cell membrane.

A true cell could be formed much later by the flattening and folding of the vesicle and the fusion of its edges to trap the chromosomes and ribosomes inside. If this turning "outside-in" to make a true cell happened only after the evolution of membrane transport proteins, which would allow nucleotide

The procaryotic bacterial cell has no membrane-bound nucleus but a coiled ball of DNA. In most bacteria the equivalent of the mitochondria in eucaryotic cells is the mesosome, the site of cell respiration, which is formed by infolding of the cell wall. The cytoplasm contains ribosomes, enzymes and sometimes glycogen granules, or food stores. Flagella aid movement of the cell, which is usually covered by a layer of slime.

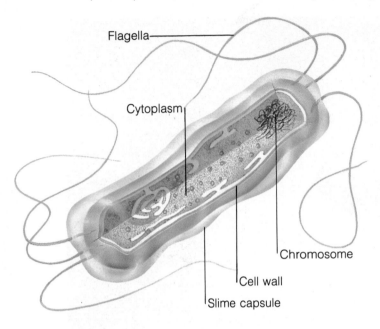

Flagella

Cytoplasm

Chromosome

Cell wall

Slime capsule

have nucleosomes or nuclear membranes, and do not divide by mitosis. Eucaryote nuclei are always bounded by a double-membraned nuclear envelope and divide by mitosis. They have a contractile machinery and a cytoskeleton consisting of microtubules. Microfilaments containing actin, together with myosin (akin to the protein of the same name that occurs in muscle) provide the contractile machinery able to move things about in all eucaryotic cells.

Microfilaments and microtubules are never found in procaryotic cells (other than as actin gene sequences), in which molecules move about the cytoplasm only by diffusion — in eucaryote cells substances can be actively pushed or pulled. The origin of the eucaryotic cell probably involved the origin of this contractile machinery.

Other major innovations for eucaryotes are the endoplasmic reticulum, the Golgi apparatus, and lysosomes — three main membrane-bound components of eucaryotic cell structure. None of these, nor the processes of taking in food particles (endocytosis) or getting rid of wastes by exocytosis, are ever found in procaryotes. Such structures and processes are essential for the secretion of proteins by eucaryotic cells, but bacteria carry out secretion much more simply. Proteins pass straight out of the cell, and so no endoplasmic reticulum or Golgi apparatus is necessary.

Another important difference is that procaryotic cells lack sex. In eucaryotes reproduction is by the fusion of eggs and sperm (or simpler gamete cells in lower eucaryotes). Moreover, procaryotic chromosomes consist of circular, rather than linear, DNA molecules and lack the histone proteins associated with DNA in eucaryotic chromatin. They also lack nucleoli and never have mitochondria or chloroplasts. The hairlike flagella of bacteria are also totally different from those of eucaryotes. So how did these dramatic differences arise?

and amino acid molecules into the cell, the cell would be able to survive. Blobel's folding mechanism would make a cell bounded not by a single plasma membrane but by two concentric ones. This apparently complex arrangement is found today in most eubacteria, including the green and purple bacteria and the cyanobacteria, and can explain the structures of a number of cell organelles.

It has been suggested that strengthening of the cell wall, as well as a DNA chromosome and a primitive division mechanism, all evolved during the inside-out cell stage. If they also became associated with pigment vesicles containing chlorophyll and evolved a type of primitive anaerobic photosynthesis, then the inside-out cell would have an abundant source of energy.

Following this scenario, it would appear that virtually all the major steps in the formation of a cell occurred in the "inside-out" cell. Then when such an inside-out cell folded outside-in to make a true cell, the result would have been a primitive green sulfur bacterium.

Procaryote versus Eucaryote

The difference between the cell of the procaryotic bacteria and the eucaryote cells of higher organisms is the most profound and important one in the whole of the living world — the procaryotes never

Origin of the Eucaryote Cytoskeleton

In 1970 it was proposed that what triggered the evolution of the much more complex eucaryotic cell was the development of the capacity to engulf other substances by the process called phagocytosis and to digest them internally. (No bacteria can feed like animals by engulfing other cells.) The first step

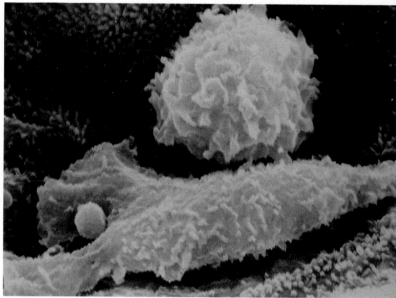

would have been the loss of the bacterial wall. The bacterial ancestor probably secreted enzymes into its environment to digest external food. As in all bacteria, its digestive enzymes would have been made by ribosomes on the inside of the plasma membrane, and passed directly to the exterior.

It must have been an exceptionally large bacterium; several such exist now. Loss of the rigid wall would allow the flexible plasma membrane to partly wrap round smaller bacterial prey. Incomplete encirclement of the prey would make digestion more economical, because secreted digestive enzymes would be less likely to diffuse away and be wasted; it would be the first step in the evolution of phagocytosis.

Loss of the cell wall would also weaken a large bacterium but the evolution of an internal meshwork of actin microfilaments attached to the plasma membrane, and cross-linked to each other, would strengthen it and prevent it from bursting. Thus might have been born a primitive eucaryote cytoskeleton — never found in any bacterium. For the cell to grow, however, it must be able to break the cross-links between the filaments to allow expansion. Such breaking and remaking of the links switches the cytoplasm to and fro between a liquid "sol" state and a jellylike or "gel" state.

Reversible sol to gel switching is essential for ameboid movement (locomotion by means of pseudopodia, temporary localized protrusions of the cell membrane and cytoplasm), common in eucaryotes, but absent in bacteria. Also needed for such movement is the protein myosin and a method of splitting the energy-rich phosphate ATP, providing the motile force by pushing against actin, as in a muscle. Bacteria use such ATP-splitting proteins to pull apart the two strands of the DNA double helix during replication. Possibly a gene for one of these was duplicated, one copy being conserved in its old form for further replication while the other mutated to become a primitive myosin molecule.

Throughout evolution, there are numerous known examples of gene duplication followed by divergence of two copies; it is the basic way of increasing the number of different available genes as organisms become more and more complex. A well-known example is in the evolution of human hemoglobin, the oxygen-carrying pigment which gives blood its red color. Humans have eight different genes coding for different forms of the protein globin, the sequences of which reveal that they all evolved from a common ancestor by gene duplications. By determining which other vertebrates have which gene, scientists can give approximate dates to the gene duplications.

Fundamental Eucaryote Features

The evolution of the ability to contract the cytoplasm had several important consequences. It allowed the cell to creep by ameboid motion in search of prey and away from danger. It was used to constrict the cell into two daughters, and became part of the mechanism of cell division. But above all it could be used to pull inside the cell the portion of plasma membrane to which prey had become stuck so as to make a self-contained food cavity or vacuole. Phagocytosis could now occur, and internal membranes would be available for the origin of the nuclear envelope, endoplasmic reticulum and Golgi apparatus.

Initially, the food vacuole would have ribosomes on its "outer" (cytoplasmic) surface and so be a primitive form of endoplasmic reticulum; when the food was digested, the cells would be filled with vesicles covered in ribosomes. Such vesicles would be moved around the cell by actin and myosin. If the cell was feeding rapidly, it might soon run out of plasma membrane and end up with too many of these internal vesicles, which would therefore be more prone to fuse with the plasma membrane. Mutations making such membrane fusion more efficient would be advantageous, and the mechanism of exocytosis (the secretion of substances to the exterior via a vesicle) could thus have come into being.

The mechanism of replication gives a clue to the origin of the nucleus. In bacteria, DNA is attached directly to the plasma membrane. As they replicate, the two daughter DNA molecules are separated from each other by the growth of the membrane and cell wall between their two attachment sites. This simple mechanism would have been present in the ancestor of eucaryotes. However, when the ancestor lost its rigid wall, separation would be less efficient and frequent mistakes could kill one or both daughter cells. Mutations producing a new rigid structure that would help to control the separation process and make it more accurate would therefore be beneficial.

In eucaryotes, control of the DNA separation is by the rigid spindle microtubules. Their origin must have been an early event in the origin of the eucaryotic cell. It is likely that phagocytosis would pull inside the cell bits of membrane attached to the

DNA, so that the DNA would no longer be attached to the cell surface but to internal membranes. These could fuse with the endoplasmic reticulum membranes to form the nuclear envelope.

Early exocytosis was probably the indiscriminate fusion of any internal vacuole with the plasma membrane. But it would be wasteful for food vacuoles to fuse with the plasma membrane before their contents were digested. It would also be wasteful for internal rough endoplasmic vesicles (which make digestive enzymes) to fuse with the plasma membrane instead of with food vacuoles. Cells best suited to survival, therefore, would be those that could differentiate between plasma membrane and food vacuoles, probably by inserting a "label" protein into the food vacuole membrane.

Similarly advantageous was the evolution of two separate types of vesicle: one containing digestive enzymes that would fuse with food vacuoles (these would be primitive lysosomes), and one carrying substances destined for the plasma membranes. So as not to waste digestive enzymes, those destined for the plasma membrane would have to be smooth vesicles, lacking ribosomes. Eucaryotic cells have a mechanism to bud off such smooth vesicles from the rough endoplasmic reticulum, and as soon as it evolved ribosomes would no longer be carried to the plasma membrane.

It is clear that all the fundamental and universal features of eucaryotes could have evolved as the natural consequence of the loss of cell walls and the evolution of phagocytosis. But the scale and complexity of the changes are so great that success in the early stages seems improbable. This is also suggested by the late stage at which they occurred — 1,500 million years ago, 2,000 million years after the origin of life. Had they been probable processes, they would surely have occurred earlier.

The Creation of Chloroplasts

Eighty years ago, the Russian Mereschkowsky made the revolutionary proposal that chloroplasts (the chlorophyll-containing bodies in the cells of green plants) evolved from cyanobacteria that had been taken in by a protozoan cell and not digested. Instead they became useful partners — symbionts — and then transformed into true cell organelles. It

Giant algal cells in the swamp waters of Florida resemble the heads of parasols. The green color of these cells results from the predominance of chloroplasts containing the photosynthetic pigment chlorophyll.

is now known that, like cyanobacteria, chloroplasts are bounded by an envelope consisting of two membranes and they contain circular DNA and ribosomes, all thought to be derived from the cyanobacterium rather than the host.

Evidence for Mereschkowsky's symbiosis theory comes from the blue-green chloroplasts of the unusual unicellular eucaryote alga called *Cyanophora*. The envelope surrounding these chloroplasts resembles a cyanobacterial cell wall, and initially the chloroplast was thought simply to be a symbiotic cyanobacterium. Then the DNA from the blue-green chloroplast was extracted and found to be a very short strand, as in other chloroplasts; in contrast, cyanobacteria have thousands of genes. The membrane structure shows that the chloroplast must once have been a cyanobacterium, but the DNA confirms it to be a true chloroplast. It is possible to argue, then, that the genes must therefore have been transferred from the symbiont into the host nucleus.

The problem with this theory is not how gene transfer might have happened, for genes could easily escape from a broken chloroplast and become spliced into nuclear DNA. Rather it is how chloroplast proteins coded by them (which would thereafter be made outside the chloroplast instead of inside it) could get into the chloroplast. A new transport mechanism would have had to evolve to move them across the chloroplast envelope.

This would make it improbable that the different colored chloroplasts of the green plants, red algae, brown algae, yellowish algae, and blue-green ones such as *Cyanophora*, resulted from the totally independent engulfing of different colored bacteria, as Mereschkowsky had suggested. It would seem more likely that the various pigments of the different algae might have been produced by divergent evolution from a single ancestor.

The problem of how hundreds of genes might be transferred from a symbiont to a nucleus was simplified by the discovery that virtually all organisms contain a variety of "jumping genes," which can occasionally move from one DNA molecule to another. Jumping genes code for proteins that can cut them out of one DNA molecule and insert them into another without causing any permanent damage to the DNA. Such

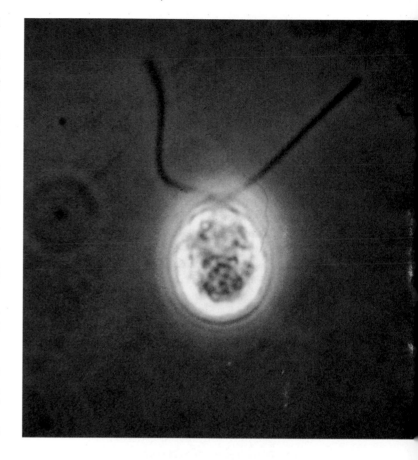

Chlamydomonas is a photosynthetic alga with flagella. By studying the effect of light on this organism, scientists have been able to work out the pathways of electron transport within the cell.

arrangements have probably been important in cell evolution.

A large number of genes must have moved from the chloroplast into the nucleus, although a few probably traveled the other way. A major feature of many such genes is that extra non-coding DNA sequences are often found inserted into the middle of the genes, like extra blank pages accidentally bound into the middle of a book. After such "split" genes are transcribed to make RNA, the extra meaningless sequences have to be cut out and thrown away and the cut ends replaced to make intact, functional RNA molecules.

Neither split genes nor RNA splicing were ever found in eubacteria such as the cyanobacteria, from which chloroplasts evolved. Split genes and RNA splicing probably evolved during the origin of eucaryotes. Yet chloroplasts do contain split genes. The extra non-coding pieces in chloroplast DNA may, therefore, have been derived from jumping genes that entered them from the nucleus.

Some jumping genes may be useful to the cell, but most are probably "parasitic" or "selfish" bits of DNA (rather like viruses) which have evolved within the cell as a result of a Darwinian struggle for existence between the DNA molecules themselves. Viruses may well have evolved from such bits of selfish DNA that became able to infect other cells. The DNA molecules within us are not only cooperating to build our body; some are competing with each other and with the more useful molecules.

The Evolution of Mitochondria

Human mitochondria are also thought by many scientists to be relics of symbiotic bacteria, which now provide most of our energy and are the reason we need to breathe oxygen. The symbiont was probably a purple non-sulfur bacterium or a close relative. According to this theory, our animal ancestors evolved not from a single bacterial species, but from two different types of bacteria; the one that evolved into the eucaryotic cell, and the symbiotic purple non-sulfur bacterium. But did our ancestors take up their symbiotic bacteria before or after they became eucaryotes?

Lynn Margulis of Boston University has suggested it happened before the eucaryotic cell evolved.

Many other biologists think this improbable, for present-day bacteria cannot engulf prey. More probably phagocytosis (the engulfing mechanism) happened first, and only after the eucaryote cell evolved did one of them engulf a purple non-sulfur bacterium to form a mitochondrion.

Indeed, more than a thousand species of protozoa never have mitochondria and do not need oxygen. The British biologist Tom Cavalier-Smith calls them the Archezoa and argues that many — if not all — of them may be relics of the first eucaryotes that did not have mitochondria and so did not need oxygen. Today, they survive only in anaerobic habitats, such as mud, stagnant water and inside animal intestines. This is because in an oxygen-containing environment they cannot compete with protozoa that do have mitochondria. Oxygen-based respiration by mitochondria yields several times as much energy from the same amount of food as the fermentation processes used by the Archezoa. When mitochondria evolved, the cells containing them would have displaced the archezoan protozoa from oxygen-rich environments, but not from anaerobic ones.

Purple non-sulfur bacteria are the most likely ancestors of mitochondria, because their respiratory enzymes are the most similar to those of mitochondria. They photosynthesize in light in the absence of oxygen. But in the dark and in the presence of oxygen, however, photosynthesis is suppressed and purple non-sulfur bacteria use oxygen for respiration just as mitochondria do.

Cavalier-Smith has proposed that the protozoan which engulfed the cyanobacteria which eventually became chloroplasts was also the one that engulfed purple non-sulfur bacteria to form mitochondria. The waste oxygen produced by the cyanobacterium would be used by the purple non-sulfur bacterium for respiration. Simultaneous conversion of the symbionts into chloroplasts and mitochondria would have produced the first unicellular algae, which were ancestral to multicellular algae (like modern seaweeds) and plants. Subsequent loss of chloroplasts would have yielded the non-photosynthetic protozoa that do have mitochondria, and which eventually gave rise to fungi and to the whole of the animal kingdom.

The conversion of a bacterium into a mitochondrion is similar to that for a chloroplast: massive gene transfer into the nucleus, coupled with a new mechanism to transport the proteins coded by those genes back into the mitochondrion. But in mitochondria the gene transfer went much farther than in the case of the chloroplast. Only eight to fourteen protein-coding genes remain in mitochondrial DNA in addition to those coding for mitochondrial ribosomal RNA and transfer RNA. This drastic reduction in proteins coded by mitochondrial DNA allowed its genetic code to be simplified. Consequently the mitochondrial genetic code differs somewhat from the universal code used by bacteria, chloroplasts and the nucleus.

Besides chloroplasts and mitochondria, there appears to be a third symbiosis — the unique cells of the seaweeds known as kelps. The chloroplasts of the marine kelps and other brown seaweeds do not lie free in the cytoplasm as do those of ordinary plants. Instead they are enclosed by two extra

membranes — called the chloroplast endoplasmic reticulum. These extra membranes also occur in various brownish unicellular algae, such as diatoms, which are important organisms in plankton.

The ancestors of the brown algae were probably colorless protozoa which fed by phagocytosis. One of them — or perhaps millions of them each — ate a photosynthetic eucaryote cell with brown chloroplasts, which became a symbiont (possibly because of its own rapid cell growth) and was eventually converted into the chloroplast of the host. ''Missing links'' that strongly support this theory are the planktonic cryptomonads. They are simple cells with two cilia clothed in curious tubular hairs, like those on one of the two cilia of the sperm of brown algae and the water molds known as Oomycetes. The distinctive cell structures shown by these organisms have recently provoked the suggestion that they be classified into a kingdom of their own, the Chromista, separate from the other four eucaryote kingdoms: plants, animals, fungi and protozoa.

Multicellular Structure

Even after all the evolutionary processes described so far — which took around 3 billion years — life

The intricate patterning of diatoms
(below and left) *makes them the*
architectural marvels of the sea.
These single-celled algae are
important producers of oxygen by
photosynthesis in the oceans.

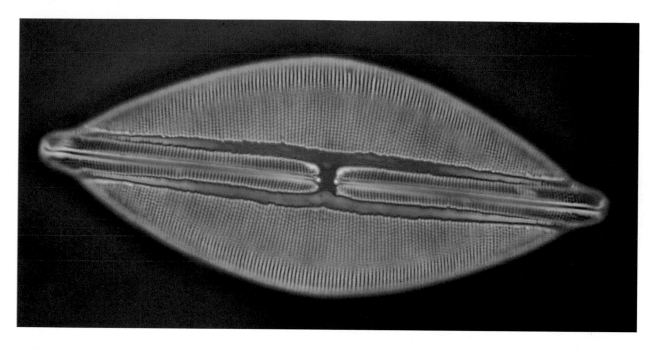

still took the form of only single-celled organisms. The next important step was the development of multicellular organisms and then the various body-plans that characterize higher plants and animals

Becoming multicellular can be regarded as easy, requiring only adhesives — gums in plants and fungi, and glues in animals — to stick the cells together. But for many organisms, there are no advantages in doing so.

In the open ocean, a photosynthetic cell is continually tumbled around by water currents. There would be no advantage in growing big and developing leaves and roots. Each cell can get its minerals, light and water much more efficiently if it stays separate and is not trapped deep inside a multicellular body.

Conditions are very different, however, on the seashore and on land. An algal cell attached to a rock near the shore can grow faster than in the open ocean, because coastal waters are continually fertilized by minerals washed from the land and supplied by upwelling currents from the seabed. And being fixed to a rock they are not washed away into the comparative desert of the open sea.

For attached cells, however, the energy-giving light comes always from above, not from all sides as with the tumbling planktonic cell. Any other cells

growing taller overshadow it and starve it of light. An upward struggle commences. Mutations that increase its size help the organism to survive. Hence, becoming multicellular could have been a means of becoming larger (taller) and competing for light. This is probably why attached seaweeds near the shore are multicellular, whereas the free-floating plankton in the ocean is unicellular. On the seashore, a multicellular structure is evidently so advantageous that it evolved in three separate groups: the red, green and brown seaweeds.

Fungi on land live on rotting vegetable or animal matter. They feed by penetrating organic food with millions of microscopic threads (hyphae). For secreting digestive enzymes into their environment and absorbing the digested food, they need a large surface area. The hypha of a feeding fungus must therefore remain microscopically thin. It can grow very long, but never fat, or it would lose out to thinner rivals.

A large fungus body, such as a mushroom, is never the feeding stage. Fungal hyphae collaborate to make massive bodies only for dispersing their reproductive spores, which travel on air currents. Wind blows faster above the ground than close to it. To travel farthest, a fungus spore must be thrown high, so mushrooms grow on stalks — the higher

the better. Spores must also be protected from rain, which could batter them to the ground before they travel far, so the umbrella shape of a mushroom protects its spores from the rain.

The true fungi evolved from a protozoan cell with a single cilium, which grew a cell wall and became a saprophyte. The oomycete water molds, or pseudofungi, evolved entirely independently from a brownish alga that lost its chloroplasts.

Among the simplest multicellular animals are the sponges which live attached to underwater rocks. They have no gut or true nervous system, but are not merely a large colony of protozoan cells. They have more than a dozen different types of cell, most of which are naturally found in higher animals. As do other animals, sponges have flat sheets of cells (epithelia) closely cemented together, and more independent connective tissue cells embedded in a jellylike extracellular material containing collagen — a primitive skeleton.

Thus sponges are excellent "missing links" between protozoa and animals that have a proper nervous system and a gut. They feed by filtering seawater (only a few live in freshwater) through channels in their body to trap unicellular prey, mainly bacteria, which are then engulfed and digested by the cells lining the water channels. The water is driven through the channels by the beating of hairlike flagella and the bacteria are filtered from it by a tenuous collar of minute projections (microvilli) that surround each flagellum.

A fascinating group of protozoa consists of single collar cells, or colonies of them, which feed in exactly the same way as sponges. Like them, they live attached to solid objects. One most probably evolved into the first sponge by multiplying its cells to make a multicellular "net" able to catch far more bacteria. The bigger the net, the more necessary were other non-ciliated cells to support it and raise it farther off the seabed. The evolution of muscles probably followed, for all protozoa have the contractile muscle proteins actin and myosin. Thus the first multicellular animals fed as protozoa by filter-feeding and phagocytosis.

The Nervous System and Higher Animals

The simplest animals with a nervous system are the cnidarians — jellyfish, sea anemones, corals and

hydroids. Virtually all of these organisms have a polyp (nonmotile) stage stuck on to a rock, which a minority of zoologists think evolved from the simplest calcareous sponges. The central water-filled cavity of the sponge would become the gut cavity, and its major opening would become the mouth of the polyp. The biggest change, besides the origin of the nerves, would be the origin of the stinging cells, the cnidocytes, which give the cnidarians their name.

Cnidocytes contain explosive capsules called nematocysts. On suitable stimulation, these evert stinging threads from the surface of the cell. They form the stings of jellyfish and sea anemones, some of which can be severely poisonous to humans — for example, the notorious stings of the Portuguese man o' war. The cnidarians use them not only for stinging and paralyzing prey, but for a variety of other purposes, including attachment. Their whole way of life centers on their nematocysts — a specialized protein secretion of the endoplasmic reticulum. The origin of nematocysts may, therefore, have been the key step in the conversion of a sponge into a cnidarian.

Nematocysts and similar explosive capsules have evolved several times, apparently independently, in several groups of protozoa. It appears, therefore, that they evolve easily. And according to the minority view whereby polyps evolved from the simplest calcareous sponges, it is likely that a sponge evolved nematocysts, which enabled it to catch larger prey than its existing collar cells, and so became the first corallike polyp. The jellyfish stage found in many cnidarian life cycles would have evolved later by the budding off from the polyp of a free-floating dispersal stage.

Sponges have no nerves, and for this reason most zoologists think that they could not have given rise to higher animals in this way. Instead they imagine some kind of free-swimming ancestor. But the subject is controversial and there is not yet any general agreement about the identity of the ancestor of higher animals.

In fact, sponges contain a network of branching cells that resemble nerve cells and which contain the chemical transmitters used to send nerve signals from cell to cell. If these cells also evolved the capacity to conduct electrical impulses, they could become true nerve cells. One group of sponges, the deep-sea glass sponges, has even evolved electrical impulses; but their body structure is so different from that of all other animals that many commentators find it reason enough to believe that they could not be our ancestors — instead of being divided into separate cells, they have numerous nuclei in a common cytoplasm.

Exploration of the ocean's depths has revealed a host of animals of extraordinary shapes. The evolutionary history that led to the variety of forms may be proved only by further genetic studies of these animals, many of which live in total darkness and at tremendous pressures.

The evolution of electrical impulses is not difficult. Every plasma membrane has proteins that pump ions across it, making it electrically polarized. And nervelike electrical impulses have evolved in several different protozoa and plants.

The origins of organisms that have both a mouth and an anus, and a left and a right side is even more controversial. Some zoologists regard some kind of creeping worm as most primitive, whereas others favor the idea of an attached filter-feeder, such as the bryozoa or "moss animalcules."

Are the attached filter-feeders such as barnacles, bryozoa, mussels and tube-worms primitive? Or did they evolve from mobile relatives that settled down and lost their head? Are flatworms primitive because they all have no anus? Or did their ancestor lose its anus, as has clearly happened to other simple animals? Zoologists disagree about the answers to these questions. Perhaps one day molecular studies of the genes will resolve the controversy, but it will be necessary to study not just the familiar animals around us. The sedentary polyplike animals must not be forgotten, nor must those that live in bizarre habitats, such as the mid-oceanic rifts now being explored by submersibles.

Alternative Energy Sources

Deep down on the ocean floor near the mid-ocean rifts, where light never penetrates, remarkably rich animal communities have been discovered recently alongside the volcanic cracks from which sulfurous hot water pours. They depend not on plants trapping the energy of sunlight, but on bacteria that get their energy from hydrogen sulfide — a pungent gas. The animals in these "oases" of life are all blind. Most striking is a giant tubeworm, five feet tall. This creature, named *Riftia*, is a member of the phylum Pogonophora which have virtually all lost their gut. Instead they have a mass of springy tissue in which they cultivate millions of the bacteria that feed on the hydrogen sulfide and provide them with their food. Chemosynthetic bacteria have long been known, but this is the first discovery of an entire ecosystem totally dependent on them.

Is light the only long-term energy source that could support life? Or could a planet orbiting a dead sun evolve life based on chemosynthesis and the heat generated by its own radioactive decay? Probably not, for the reactions that chemosynthetic bacteria use to get energy to make organic molecules require not only hydrogen sulfide but also oxygen from the surrounding seawater — and that was created by photosynthesis. So the *Riftia* communities are independent of light only in the short term. They could not have evolved until after photosynthesis oxidized the atmosphere. Perhaps, therefore, life elsewhere in the universe is also based on photosynthesis.

It is sometimes suggested that life on other planets could be based on silicon instead of carbon. But amino acids and nucleic acid bases form so very easily in reducing atmospheres, and many of their constituent elements are far more common in the universe than is silicon, which does not form a similar set of molecules. For these reasons it is likely that other life in the universe is carbon-based and involves nucleic acids and proteins, as it does on Earth.

Whether life does exist elsewhere depends on the number of planets that are potentially suitable, and on the probability of life-forms evolving. Unfortunately, there is uncertainty on both counts. There is no sound evidence for life elsewhere in the Solar System. Increasingly, however, astronomers think that vast numbers of planets probably exist outside the Solar System, and there is some evidence for planets around nearer stars. Furthermore, molecules essential for the formation of life have been detected throughout the universe by radioastronomers, although life itself has not yet been found.

Many biologists think life would inevitably evolve given suitable conditions, but until a better understanding is reached of how the genetic code evolved, this also must remain in doubt. Nonetheless, plans are well advanced for a massive listening program by radio telescopes in an attempt to detect intelligent life elsewhere in our galaxy. The chance of success depends on the probability of any such beings surviving for sufficient time for us to make contact, as well as on the probability of intelligence evolving in the first place. Given a eucaryotic cell, the eventual evolution of animals and a nervous system might be inevitable. But, as this chapter has shown, the evolution of the eucaryotic cell was so

The stupendous kaleidoscope of colors and the abundance of species in the sea is a constant marvel, not only because of the spectacle itself but also because life on Earth probably first evolved in the sea.

145

The development of space technology
has enabled man to probe into the
universe and discover more about the
evolution of life. Radiotelescopes
(above) track the movement of
objects and record events in the sky,
and space rockets and the astronauts
who accompany them carry out
research and investigations in space.
Spectroscopy has shown that the
same atoms and primitive molecules
are found throughout the universe,
which may make it possible for life to
have developed elsewhere.

146

difficult and improbable — and took so long — that it is certain to occur differently, if at all, were evolution to start again. And even given a nervous system, the evolution of intelligence does not necessarily seem inevitable.

The British molecular biologist Francis Crick and his colleague Leslie Orgel suggested that intelligent life elsewhere might deliberately have seeded the Earth with bacteria and that life (as we know it) may not have originated on Earth at all. One objection to this view, pointed out by Crick himself, is that such an outside intelligence would surely also have sent eucaryotic spores.

Other scientists, from Arrhenius in the last century to the British astronomer Fred Hoyle today, have proposed that life originated in outer space from where its spores naturally rained down on to the Earth. But to most biologists these suggestions are the themes of science fiction rather than scientific contributions to the problem of the origin of life. Even if life is found nowhere else, man can happily work to understand it and ensure its survival here on Earth.

The possibility of the existence of extraterrestrial life has been a source of inspiration for writers from H. G. Wells to the creators of contemporary sci-fi books and movies. These creatures from The Invasion of the Saucermen *were the products of fantasy, not scientific fact.*

Appendix

The Phyla of Life on Earth

About 1.5 million distinct kinds of living organisms have been described, and twice as many probably remain to be discovered. The diversity of life is so overwhelming that biologists can make sense of it only by arranging or classifying species into groups of related species; the science of doing this is called taxonomy. Those species with a similar general structure are placed in the same family group, for example, tigers, lions, lynxes and cats in the family Felidae. Related families are grouped into orders, like the Carnivora which contains dogs, cats and bears; orders are grouped into classes (e.g. Mammalia), classes into phyla, and phyla into kingdoms.

As one proceeds up the taxonomic hierarchy from family to order, class and phylum, the number of characteristics shared by a group become fewer and fewer because of evolutionary divergence, until at the level of a phylum they share only fundamental features of their body plan. Biologists use all aspects of structure — outward appearance, internal anatomy, cell and molecular structure — in classifying organisms and must take into account all stages of their life cycle. New discoveries, therefore, necessitate modifications to existing classifications to make them more accurately reflect true relationships and differences.

Expert and balanced judgment is needed to describe the relative importance of different characteristics and the correct level in the hierarchy for a particular group. Because this cannot be made totally objective, and new knowledge takes time to assimilate, different present-day classifications disagree somewhat in the number of phyla and kingdoms they recognize.

The traditional division simply into the animal and plant kingdoms was superficial. The more modern classification given here emphasizes basic cell structure. The biggest gulf in the living world, indeed, is between ordinary organisms made of cells (the Cytota), and the viruses and viroids which are noncellular and therefore placed in a separate "empire," the Acytota. For cellular organisms also the fundamental division is not between animals and plants, but between the bacteria, which have nonnucleated procaryotic cells, and the eucaryotes with nucleated eucaryotic cells.

Even for eucaryotes the traditional division into plant and animal kingdoms is misleading. Fungal cells differ even more from plant cells than they do from animal cells, so fungi must be treated as a distinct third kingdom. The same is true of the marine kelps and their relatives which have been recently recognized as a separate kingdom, the Chromista. Though the unicellular Protozoa are still sometimes left in the animal kingdom they are now more commonly removed from it and lumped together with the Chromista to form a single kingdom, the Protista or Protoctista. But this is such a heterogeneous mixture, almost impossible to define precisely, that it is preferable to treat the Protozoa as a kingdom on its own.

LIVING ORGANISMS

EMPIRE 1	CYTOTA	CELLULAR ORGANISMS
Superkingdom 1	**Procaryota**	**Cells without nuclei or mitosis**
Kingdom 1	Bacteria	Bacteria
Superkingdom 2	**Eucaryota**	**Cells with nuclei and mitosis**
Kingdom 1	Protozoa	Protozoa
Kingdom 2	Chromista	Brown seaweeds and their relatives
Kingdom 3	Plantae	Green plants and red seaweeds
Kingdom 4	Fungi	Fungi
Kingdom 5	Animalia	Animals
EMPIRE 2	ACYTOTA	NONCELLULAR ORGANISMS
Kingdom 1	Vira	Viruses and viroids

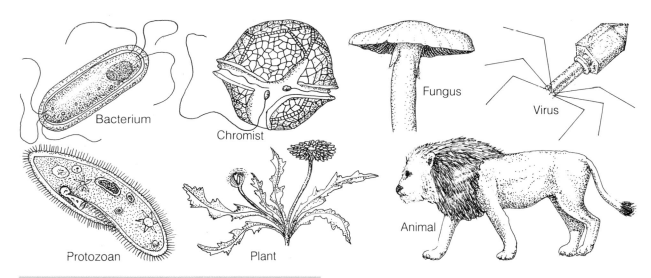

Bacterium

Chromist

Fungus

Virus

Protozoan

Plant

Animal

Kingdom Bacteria

Most bacteria are unicellular, although the blue-green cyanobacteria form multicellular filaments, and the soil-dwelling actinomycetes form long branching threads superficially resembling fungal mycelia. All bacterial cells are procaryotic: they never have nuclei, mitosis, endoplasmic reticulum, Golgi apparatus or mitochondria. The ability to contract the protoplasm, phagocytosis ("feeding" by engulfing particles) and secretion by exocytosis (vesicle fusion with the plasma membrane) are entirely absent. Bacterial flagella are much simpler than the eucaryotic cilia and flagella. Their cells are usually bounded by walls consisting of rigid peptidoglycan, or more rarely of protein.

Bacteria obtain energy in a far greater variety of ways than do eucaryotes. Some respire using oxygen, others use nitrate or sulfate instead. Some are anaerobic (oxygenless) fermenters. Some photosynthesize like plants, whereas others photosynthesize without producing oxygen. Some feed on organic compounds. Many are useful, and most harmless, to man.

Bacteria placed in the subkingdom *Muramibacteria* have muramic acid in their walls, and differ in this and many other chemical respects from those in the subkingdom *Archaebacteria*. A few bacteria in both subkingdoms have no cell walls, and so are naturally resistant to antibiotics that act by damaging bacterial cell walls.

KINGDOM BACTERIA

SUBKINGDOM 1	MURAMIBACTERIA	
Superphylum 1	**Gracilicuta**	**Typical gram-negative bacteria**
Phylum 1	Chlorobacteria	Green sulfur bacteria
Phylum 2	Rhodobacteria	Purple bacteria; gram-negative eubacteria
Phylum 3	Oxyphotobacteria	Cyanobacteria; Prochloron
Phylum 4	Spirochaetae	Spirochetes, syphilis
Phylum 5	Myxobacteria	Slime bacteria
Superphylum 2	**Unicuta**	**Gram-positive bacteria and mycoplasms**
Phylum 1	Firmicuta	Gram-positive bacteria; actinomycetes
Phylum 2	Mollicuta (= Tenericutes)	Mycoplasmas
Phylum 3	Radiobacteria	Radiation-resistant cocci
SUBKINGDOM 2	ARCHAEBACTERIA	
Phylum 1	Methanobacteria	Methane-producing bacteria
Phylum 2	Halobacteria	"Salt-loving" (halophilic) bacteria
Phylum 3	Thermacidibacteria	Bacteria liking hot acid

Kingdom Protozoa

Most protozoa are unicellular eucaryotes that usually feed by phagocytosis — engulfing their "prey". Their feeding stages lack cell walls, although resting or dispersive stages often have them. A few individual protozoan cells grow large enough to be visible to the naked eye; others produce large masses of mobile protoplasm containing numerous nuclei. Many form colonies of identical cells, or even simple multicellular organisms with two or three different cell types.

Many protozoa have a sexual reproductive stage, but their sex cells are usually not distinguishable as male sperm and female eggs. They cannot photosynthesize — except for some euglenoids and dinoflagellates; these do have chloroplasts with a bounding envelope containing three concentric membranes, not two as in Plantae and Chromista. Their mitochondria usually have tubular rather than platelike cristae.

KINGDOM PROTOZOA

SUBKINGDOM 1	ARCHAEOZOA	PROTOZOA WITH NO MITOCHONDRIA
Phylum 1	Archamebae	Mastigamebae and other amebae lacking mitochondria
Phylum 2	Metamonada	Trichomonads, hypermastigotes, oxymonads, retortamonads, diplomonads
Phylum 3	Microspora	Microsporidian parasites
SUBKINGDOM 2	MITOZOA	PROTOZOA WITH MITOCHONDRIA
Phylum 1	Euglenozoa	Euglenoids, trypanosomes, pseudociliates
Phylum 2	Dinozoa	Peridineans ("dinoflagellates"); syndineans
Phylum 3	Proterozoa	Proteromonads, opalinids, cyathobodonids, plasmodiophorans
Phylum 4	Heterocaryota	Ciliophora ("ciliates" and suctoreans)
Phylum 5	Mesozoa	Mesozoa (simple multicellular parasites)
Phylum 6	Sporozoa	Gregarines, coccidians, malarial parasites
Phylum 7	Mycetozoa	Slime molds
Phylum 8	Sarcodina	Amebae (having mitochondria); foraminifera, heliozoa, radiolaria
Phylum 9	Ascetospora	Haplosporidian parasites
Phylum 10	Myxozoa	Myxosporidian parasites
Phylum 11	Choanozoa	Choanoflagellates

Kingdom Chromista

Most chromists are unicellular or multicellular algae, with brown, yellowish-brown, or yellowish-green chloroplasts. Unlike plant and protozoan chloroplasts, they are always separated from the cytoplasm by the membranes of the chloroplast endoplasmic reticulum. Unicellular species, like the diatoms, thrive in marine and freshwater plankton; others, like the giant kelps, form the most important vegetation on the coastal seabed. Many have mobile unicellular stages with usually two cilia, one or rarely two being covered with tubular hairs — with the exception of the Haptophyta, which do not have hairs on their cilia.

The kingdom also includes a few non-photosynthetic organisms possessing the same ciliary structure: the fungus-like pseudofungi; the marine labyrinthulids; and the bicosoekids.

This kingdom includes the green plants and red seaweeds: single-celled or multicellular eucaryotes with chloroplasts bounded by a double-membrane envelope, but with no chloroplast endoplasmic reticulum. All photosynthesize, except for a few secondarily nonphotosynthetic parasites on other plants.

Starch is stored in the chloroplasts in green plants and outside the chloroplasts in the red seaweeds, which have red (and blue) pigments that usually mask their green chlorophyll. All plants have mitochondria with platelike cristae.

Yeasts are unicellular fungi. Nearly all others consist of multicellular filaments called hyphae. Yeasts never have chloroplasts nor feed by phagocytosis. Fungal hyphae are surrounded by walls of chitin (or similar macromolecules), the rigid carbohydrate that forms the "skin" of insects. They digest the organic materials in which they grow with the aid of enzymes secreted outside the hyphae, the digestion products being actively absorbed through their surface. Like animals they store carbohydrate as glycogen, and they have mitochondria with platelike cristae.

KINGDOM PLANTAE

SUBKINGDOM 1 BILIPHYTA
(ALGAE WITH PHYCOBILIN PIGMENTS)

Phylum 1	Biliphyta	Red algae (Rhodophyceae) and glaucophytes (e.g. Cyanophora)

SUBKINGDOM 2 VIRIDIPLANTAE
(GREEN PLANTS, WITHOUT PHYCOBILINS)

Phylum 1	Chlorophyta	Green algae
Phylum 2	Bryophyta	Mosses, liverworts, hornworts
Phylum 3	Tracheophyta	Vascular plants

Subphyla:
Psilophyta: Rootless plants
Lycophyta: Clubmosses
Sphenophyta: Horsetails
Pterophyta: True ferns (homosporous)
Heterosporophyta: Flowering plants, conifers, ginkgos, "water ferns," Selaginella

KINGDOM FUNGI

SUBKINGDOM 1 CILIOFUNGI
(CILIATED FUNGI)

Phylum 1	Chytridiomycota	Chytrids, various water molds

SUBKINGDOM 2 EUFUNGI
(NONCILIATED FUNGI)

Phylum 1	Ascomycota	Yeast, powdery mildews, blue molds, morrels, truffles, cup fungi
Phylum 2	Zygomycota	Black bread mold, insect-destroying fungi
Phylum 3	Ustomycota	Smuts (plant parasites)
Phylum 4	Uredomycota	Rusts (plant parasites)
Phylum 5	Basidiomycota	Mushrooms, toadstools

KINGDOM CHROMISTA

SUBKINGDOM 1 HETEROKORTA
(TUBULAR HAIRS ON ONE CILIUM ONLY)

Phylum 1	Pseudofungi	Oomycetes, hyphochytridiomycetes
Phylum 2	Chromophyta	

Subphyla:
Chrysophytina: Chrysophytes, brown seaweeds, yellowish algae, labyrinthulids
Bicosoekida: Bicosoekids
Chloromonada: Chloromonads
Silicociliata: Silicoflagellates
Bacillariophytina: Diatoms
Eustigmatophytina: Eustigs

SUBKINGDOM 2 CRYPTOPHYTA
(TUBULAR HAIRS ON BOTH CILIA)

Phylum 1	Cryptophyta	Cryptomonads

SUBKINGDOM 3 HAPTOPHYTA
(NO TUBULAR HAIRS ON CILIA)

Phylum 1	Haptophyta	Haptophyceae, cocco-lithophores

SUBKINGDOM 1	RADIATA	
Superphylum 1	**Parazoa**	**Animals with no nervous system or gut**
Phylum 1	Placozoa	*Trichoplax*
Phylum 2	Porifora	Sponges
Subphyla:	*Cellularia*	*Cellular sponges*
	Hexacta	*Syncytial glass sponges*
	Archaeocyatha	*(Extinct) archeocyathids*
Superphylum 2	**Coelenterata**	**Animals with body cavity serving as gut, no anus**
Phylum 1	Cnidaria	Corals, sea anemones, hydras, hydroids, jellyfish
Phylum 2	Ctenophora	Ctenophores
SUBKINGDOM 2	BILATERIA	
Phylum 1	Bryozoa	Polyzoa; "moss animals"
Phylum 2	Sipunculida	Peanut worms
Phylum 3	Echiuroida	Spoonworms
Phylum 4	Nemertina	Proboscis worms
Phylum 5	Platyhelminthes	Flatworms, tapeworms, flukes
Phylum 6	Gnathostomulida	Tiny worms with "micro" jaws
Phylum 7	Pogonophora	Tube-dwelling "beard worms"
Phylum 8	Annelida	Earthworms, leeches and other segmented worms
Phylum 9	Mollusca	Bivalves (clams, oysters, etc.), snails, octopus, squids
Phylum 10	Biramia	Arthropods with doubly branched limbs
Subphyla:	*Crustacea*	*Barnacles, shrimps, crabs, lobsters*
	Trilobita	*(Extinct) trilobites*
	Chelicerata	*King crabs, scorpions, spiders, mites, ticks*
Phylum 11	Entoprocta	Entoprocts
Phylum 12	Uniramia	Arthropods with branched limbs
Subphyla:	*Lobopoda*	*Onychophorans and tardigrades*
	Myriapoda	*Centipedes, millipedes*
	Insecta	*Insects*
Phylum 13	Aschelminthes	Nematode and acanthocephalan worms, rotifers, priapulids
Phylum 14	Chaetognatha	Arrow worms
Phylum 15	Phoronida	Phoronids (tube-dwelling tentacle bearers)
Phylum 16	Hemichordata	Pterobranchs, acorn worms
Phylum 17	Echinodermata	Sea lilies, sea urchins, sea cucumbers, starfish
Phylum 18	Chordata	Chordates
Subphyla:	*Cephalochordata*	*Lancelets (amphioxus)*
	Tunicata	*Sea squirts, salps, larvaceans*
	Vertebrata	*Fish, amphibians, reptiles, birds, mammals*

Kingdom Animalia

Animals are multicellular eucaryotes that feed by ingesting organic material. Their bodies consist of epithelial tissues (or sheets of cells closely cemented together), and of mesenchymal or connective tissues. Cells of connective tissue are embedded in a jellylike matrix strengthened by collagen and other fibers. Chloroplasts and rigid cell walls are absent, and phagocytosis is general. Sexual reproduction occurs in all phyla, the sex cells invariably being differentiated into large immobile eggs and mobile sperm. Typically sperm have a single posterior flagellum, but in some groups there are two, or even none. Centrioles are typically present during mitosis. All animals have mitochondria, usually with platelike cristae. All except the Placozoa have muscle cells, and all except the Parazoa have a nervous system. Adult hexact sponges differ from all other animals in being syncytial — having many nuclei in a common cytoplasm rather than being divided into separate cells.

Kingdom Vira

Viruses and viroids contain infectious pieces of DNA or RNA that can multiply only inside cells. Unlike cells they never have ribosomes or the genes coding for them, so they have to rely on the host cell for protein synthesis. Unlike other parasites they are noncellular in structure. Some species kill the host cell whereas others merely sap its energy. A virus has a coat of protein to protect its DNA or RNA; viroids have no protein coat and are such tiny pieces of RNA that they do not code for any proteins.

The virus DNA or RNA molecule (never both) acts as its chromosome, coding for its protein coat and proteins needed for its replication in the cell. The coat may be a simple icosahedral shell, a helical tube, or a more complex shape with a tail and sometimes projections like the legs of a moonlander. The chromosomes of DNA viruses and ordinary RNA viruses are replicated without a DNA intermediate. But retroviruses replicate their RNA in a complicated indirect fashion: first a DNA copy is made and inserted into the DNA of the host cell's chromosome where it can be replicated indefinitely as if it were one of the host's own genes, and also serve as a template for making new retrovirus RNA. Some DNA viruses can also insert into cellular DNA like this.

How viroids multiply is poorly understood. Apparently they are not replicated directly but made by host enzymes from a DNA template. Possibly they evolved from retroviruses by simplification and loss of protein-coding genes. Viruses are thought to have evolved from cellular nucleic acid molecules that mutated so as to allow partially independent replication and infectious transfer to new host cells. This probably happened on several occasions during evolutionary history, although it may possibly have occurred only once.

KINGDOM VIRA		
Phylum 1	Deoxyvira	DNA viruses (e.g. poxviruses, herpes and wart viruses, adenovirus, tailed bacterial viruses)
Phylum 2	Ribovira	RNA viruses (e.g. polio, common cold, yellow fever, mumps, measles, influenza, most plant viruses)
Phylum 3	Retrovira	Retroviruses (e.g. leukemia and sarcoma cancer viruses)
Phylum 4	Viroida	Viroids

Glossary

actin   a filamentous protein with a supporting function within cells, also forming, with myosin, part of a contractile mechanism that occurs widely in cells and is especially developed in muscle.

adenosine triphosphate   (ATP) a chemical which stores energy that can be released for use in the cell's metabolism.

adipocyte   a cell specialized for fat storage.

anaerobic   in the absence of oxygen.

analogy   the possession by two organisms of similar structures or attributes that have evolved separately.

anti-oxidant   a substance that prevents oxygen from combining with another substance.

antibody   a substance produced by an organism in response to foreign substances or organisms, which binds to resultant antigens and neutralizes them.

antigen   a substance which causes the production of an antibody.

ATP   adenosine triphosphate.

autophagic vacuole   a region of a cell bounded by a membrane, within which part its own substance is broken down.

bacteriophage   a virus which invades bacteria.

bilayer   a double layer of molecules; often used to refer to the structure of cell plasma membranes.

carcinogen   any substance which is able to start the formation of a cancer.

centriole   the cell organelle around which microtubules of spindle fibers are organized in animal cell division.

centromere   the part of the chromosome at which its two chromatids can be seen to be joined during cell division; the chromosome is also attached to the mitotic apparatus at the centromere.

chemotaxis   movement by an organism or cell in response to a chemical stimulus.

chemotherapy   treatment of disease by chemical methods; usually refers to the treatment of cancer by drugs that are toxic to cancer cells.

chloroplast   the organelle within a plant cell that contains chlorophyll and traps light energy.

cholesterol   a fatty alcohol found as a constituent of cell membranes, and in other animal tissues.

chromatid   one of the two threads making up the chromosomes visible at some stages of cell division.

chromatin   the combination of DNA and histone of which chromosomes are formed.

chromosome   one of the bodies within the cell nucleus visible at cell division that contains the genetic material in the form of DNA.

cilia   usually motile processes at the surface of many animal and some plant cells.

cisternal space   the part of the cell that can be regarded as inside the membrane-bounded sacs of the endoplasmic reticulum, and within which proteins may be temporarily stored before export.

clone   to produce a group of genetically identical individuals by nonsexual methods of propagation.

collagen   a protein often forming fibers; part of a body's supporting connective tissue.

condenser   a device to concentrate light on an object viewed through a microscope.

cristae   folds in the inner membrane of a mitochondrion.

cytoplasm   all of the contents of a cell apart from the nucleus.

cytoskeleton   the supporting framework of a cell, made up of protein filaments.

cytosol   fluid part of the cytoplasm.

daughter cells   two cells which are the first generation product of a cell division.

differentiation   the formation of cells or tissues with specialized functions from originally unspecialized ancestors.

DNA   deoxyribonucleic acid, a complex molecule that carries genetic information.

dynein arms   crosspieces between the microtubules within cilia and flagella, of importance in their movements.

endocrine   describes glands, or their secretions, where secretion takes place directly into the bloodstream, rather than through ducts; for example, the adrenal glands and gonads.

endocytosis   the process by which a cell takes in substances by engulfment.

endoplasmic reticulum   a system of membranous sacs within the cell which is the site of the synthesis of complex molecules.

endotoxin   a toxin produced by a bacterium, only being released when the bacterial cell wall breaks down.

enzyme   a protein which mediates in a chemical reaction within a body or cell, facilitating the reaction but, like a catalyst, not itself being consumed.

erythrocyte   a red blood cell.

erythropoietin   a hormone controlling the production of red blood cells.

eucaryote   an organism made of cells which contain nuclei; also spelled eukaryote.

exocrine   describes glands from which secretion takes place through ducts; for example, sweat glands.

exocytosis   the extrusion from a cell of cellular products packaged in a membrane.

exotoxin   a toxin secreted by a bacterium to its exterior.

flagellum   a long whiplike contractile process of a cell or unicellular organism.

flora   the plant life in a particular location; also the normal collection of bacteria that can be found on a healthy body or in the gut.

free radical an atom or group of atoms which normally exists in combination with others in a compound, but is temporarily "free" and therefore highly reactive.

genetic code the triplets of nucleotides along a DNA molecule that code for the sequences of amino acids that make up the cell's proteins.

gerontologist a scientist who studies the process of aging.

glycogen a carbohydrate storage product in cells, especially of animals; abundant in liver and muscle cells.

glycoprotein a compound of a protein and a carbohydrate; glycoproteins may cover cell surfaces.

Golgi apparatus a complex of closed membrane sacs, usually found close to the nucleus and associated with a centriole in animal cells; it plays a role in packaging and secretion of proteins and other cell products.

hematopoietic concerned in producing new blood cells.

hepatocyte a liver cell.

histology the study by microscopy of the arrangement and form of cells in tissues.

histone one of a group of proteins containing many amino acids, which binds to DNA to form chromatin.

homology a similarity of structure or other attribute in two types of organism which is the result of shared ancestry; also used to describe the relation of different amino acid or nucleotide sequences to each other.

hyaluronidase an enzyme that attacks the hyaluronate molecules that are part of the extracellular matrix.

hypha one of the minute thread-shaped parts making up the body of many fungi.

immunotherapy the treatment of disease by the activation of the body's natural defense systems.

interferon a protein which is made as part of an animal's defense against some types of viral attack.

in vitro in a laboratory apparatus rather than in a living organism.

isotope one of two or more types of atom of the same element which are identical in chemical properties but have different atomic weights; radioactive isotopes exist for many elements that make up living things.

keratin a filamentous protein which makes up hair, horn and nails, and is also found in other tissues.

lipids substances poorly soluble in water, found in living tissue, including fats stored in cells (triglycerides), waxes and allied substances, as well as phospholipids of cell plasma membranes.

lipopolysaccharide a compound of lipid, sugar and protein; part of some bacterial cell walls.

lymphocyte one of three main types of white blood cell.

lysosome a small vesicle derived from the Golgi apparatus that contains enzymes which can break down materials taken into the cell.

meiosis the type of cell division taking place during the formation of sex cells, in which pairs of chromosomes separate into different haploid germ cells.

messenger RNA a type of ribonucleic acid made using DNA as a template, which takes the code for making a protein from DNA in the nucleus to the site of synthesis at ribosomes in the cytoplasm.

metachronal rhythm the wave motion produced when rows of successive cilia beat just out of phase.

metastasis the movement of cancer cells from their original site to another part of the body.

microincineration a technique used to prepare microscopic specimens in which tissue to be studied is incinerated on a slide so as to leave only its mineral content.

microfilament a protein filament within the cell composed largely of actin and having a role in controlling cell shape and motility.

micrograph a permanent image produced through a microscope.

micromanipulation operating under the microscope with instruments such as tiny needles and pipettes on living cells.

microtome a device for cutting very thin sections of tissue to view under the microscope, or to analyze chemically.

microtubule small, intracellular tubular filaments composed of protein; they play a major role in the formation of mitotic spindles and organelles such as cilia.

mitochondrion a cell organelle concerned with the production of the cell's energy; ATP is processed in mitochondria.

mitosis the "ordinary" or "somatic" type of cell division taking place without reduction in the number of chromosomes.

mRNA messenger RNA

myelin the spirally wrapped set of membranes of Schwann cells, which form a sheath around many nerve fibers.

myosin a protein that together with actin is a major component of cellular contractile mechanisms.

nematocyst a specialized cell found in sea anemones, jellyfish and hydra, that can fire a thread to sting or capture prey.

neurotransmitter a chemical released from a nerve ending to transmit the effect of an impulse to the next nerve.

nucleolus a recognizable region within the nucleus, controlling RNA synthesis.

nucleoplasm the fluid within the nucleus; it is separated from the cytoplasm by a nuclear envelope.

objective the lens of a microscope which is nearest to the object being viewed, and forms the primary image, which is then magnified further.

oncogene a gene that leads to the formation of a tumor.

organelle a recognizable, usually membrane-bound part of a cell, which has a specific function.

pandemic a worldwide epidemic.

pathogen a microorganism that can cause disease.

peroxisome an organelle concerned with breaking down organic molecules with the use of oxygen.

phage bacteriophage.

phagocytic describes a cell able to engulf another cell or a foreign body.

phospholipid a type of fat containing a phosphate group and an organic base, as well as glycerol and two fatty acids; phospholipids have an important role in cell membranes.

photoreceptor a cell specialized as a light sensor.

photosynthesis the building-up of simple carbohydrates in plant cells using energy derived from sunlight.

pinocytosis ingestion of fluid into the cell as a droplet bounded by membrane.

plasma membrane the semipermeable membrane surrounding the protoplasm of the cell.

prebiotic before the existence of life.

procaryote an organism whose cells have no recognizable nucleus; also spelled prokaryote.

protein one of the group of compounds found in living tissue and giving it its characteristic structure. Proteins are made up of amino acids and are very large molecules; they can be chemically bonded to lipid, carbohydrate or phosphate residues, to form lipoproteins, glycoproteins or phosphoproteins, respectively.

reduced with oxygen removed, or having undergone a process in which an electron is added to an atom or ion.

rhodopsin a purple protein pigment which can be bleached by light and plays a major role in vision.

ribosomes small organelles made of RNA and protein, which may be free in the cytoplasm, or on the surface of rough endoplasmic reticulum, and are the site of mRNA bonding and protein synthesis.

spindle a system of filaments visible during cell division on which the chromosomes are arranged and moved apart.

stain a pigment or chemical with the property of coloring tissue on a

microscope slide to make it easier to see.

steroid hormone a hormone based on molecules resembling cholesterol; for example human sex hormones.

symbiosis a partnership of two organisms (symbionts) with benefits for both.

toxoid a toxin made harmless but with the capacity to provoke antibody production.

transfer RNA (tRNA) a type of ribonucleic acid that carries amino acids to the ribosomes and matches up with the correct base sequence of mRNA for protein synthesis.

vacuole a membrane-bound cavity in the cytoplasm, usually filled with fluid.

vesicle a small sac within a cell, bounded by a membrane and containing proteins or other inclusions.

virus an ultramicroscopic organism containing nucleic acid and capable of replicating itself, but only within the body of a host organism. Some viruses cause disease, many do not.

zymogen a molecule which is the inactive precursor of an enzyme and can be stored or transported without causing damage.

Illustration Credits

Life's Little Secrets

6, Eric Gravé/Science Photo Library.

The Search for Life

8, *The Creation*, mosaic from San Marco, Venice/Alinari. 10, Mary Evans Picture Library. 11, Musée Condé, Chantilly/ Bridgeman Art Library. 12, Ann Ronan Picture Library. 13, (left) Mary Evans Picture Libary, (right) **Mick Saunders**. 14, *The School of Athens* by Raphael. Vatican, Rome/Bridgeman Art Library. 15, BBC Hulton Picture Library. 18, *Christina of Sweden and her Court* (detail) by Pierre Dumesnil. Versailles/Lauros-Giraudon/ Bridgeman Art Library. 19, (top) Mary Evans Picture Library, (bottom) Ann Ronan Picture Library. 20–22, Mary Evans Picture Library. 23, (left) By courtesy of the Trustees of The National Portrait Gallery, London, (right) **Mick Gillah**. 24, (left) **Mick Gillah**, (right) Biophoto Associates. 25, Russ Kinne/ Science Photo Library. 26, **Mick Gillah**. 27, Michael Macintyre/Camerapix Hutchison. 28, Dr. G. Bredberg/Science Photo Library. 29, (left) E. H. Cook/ Science Photo Library, (right) Alvis Upitis/The Image Bank.

Marvels of a Miniature World

30, Dr. Keith Porter/Science Photo Library. 32, Eric Gravé/Science Photo Library. 33, Musée Condé, Chantilly/ Bridgeman Art Library. 34, **Mick Gillah**. 35, Dr. Brian Eyden/Science Photo Library. 36, **Mick Gillah**. 37, National Institutes of Health/Science Photo Library. 38, (left) **Mick Gillah**, (right) Dr. Gopal Murti/Science Photo Library. 39, *The Toilet of Venus* by Diego Velasquez. Reproduced by courtesy of the Trustees, The National Gallery, London. 40, Biophoto Associates. 41, Reproduced by permission of Blackie & Son Limited/ British Library. 42, (left) Division of Computer Research and Technology, National Institutes of Health/Science Photo Library, (right) Biophoto Associates. 43, *Composition for the Constructors* (1950) by Fernand Léger. Collection of M. & Mme. Maeght, Paris/ Edimedia. © DACS 1985. 45, Dr. A. Leipins/Science Photo Library. 46, Biology Media/Science Photo Library. 47, *Combing the Hair* by Edgar Degas. Reproduced by courtesy of the Trustees, The National Gallery, London. 48, Biophoto Associates. 49, **Greensmith Associates**.

The Specialized Cell

50, *Ferdinand Lured by Ariel* by Sir John Everett Millais. The Makins Collection/ Bridgeman Art Library. 52, John Durham/ Science Photo Library. 53, **Mick Gillah, Aziz Khan.** 54, Mary Evans Picture Library. 55, Michael Abbey/Science Photo Library. 56, Biophoto Associates. 57, (top) Tony Stone Worldwide, (bottom) Eric Gravé/Science Photo Library. 58, **Mick Gillah**. 59, Dr. A. Leipins/Science Photo Library. 60, (top) Dr. Gopal Murti/Science Photo Library, (bottom) John Walsh/ Science Photo Library. 61, Dr. G. Bredberg/Science Photo Library. 62, (top) Lawrence Fried/The Image Bank, (bottom) **Mick Gillah**. 63, ZEFA (UK) Limited. 64, Biophoto Associates. 65, *Daniel Lambert* by Benjamin Marshall. Leicestershire Museums, Art Galleries and Records Service. 66, *Tondo: The Virgin and Child with Saint John and an Angel* from the Studio of Botticelli. Reproduced by courtesy of the Trustees, The National Gallery, London. 67, **Mick Gillah**. 68, Michael Abbey/Science Photo Library. 69, Tony Stone Worldwide.

How Life Grows

70, J. Stevenson/Science Photo Library. 72, John Walsh/Science Photo Library. 73, Dr. Gopal Murti/Science Photo Library. 74–75, **Mick Gillah**. 76, Eric Gravé/ Science Photo Library. 77, **Greensmith Associates**. 78, Omikron/Science Photo Library. 79, *An Allegory of Prudence* by Titian. Reproduced by courtesy of the Trustees, The National Gallery, London. 80, Hank Morgan/Science Photo Library. 81, Yale University Library. 82, *The Cornfield* by Vincent Van Gogh. Reproduced by courtesy of the Trustees, The National Gallery, London. 83, *Old Women* by Francisco Goya. Musée des Beaux Arts, Lille/Bridgeman Art Library. 84, Dr. A. Leipins/Science Photo Library. 85, The Mansell Collection. 86, Prof. M. A. Epstein/Science Photo Library. 87, The Mansell Collection. 88, (top) U.S. Navy/ Science Photo Library, (bottom) Michael Friedel/The Image Bank. 89, **Mick Saunders**. 91, (top) **Mick Saunders**, (bottom) Hank Morgan/Science Photo Library.

Attackers and Destroyers

93, Österreichische Nationalbibliothek, Vienna/Bridgeman Art Library. 94, Mary Evans Picture Library. 95, *Awaiting Admission to the Casual Ward* by Luke Fildes. Royal Holloway College, University of London/Bridgeman Art Library. 96, Mary Evans Picture Library. 97, (top) Mary Evans Picture Library, (bottom) Gene Cox/Science Photo Library. 98, (top) **Aziz Khan**, (bottom) Ann Ronan Picture Library. 99, Erik L. Simmons/The Image Bank. 100–101, Ann Ronan Picture Library. 102, *Florence Nightingale at Scutari* by Jerry Barrett. By courtesy of the Trustees of The National Portrait Gallery, London. 103, **Aziz Khan**. 104, (left) Eric Gravé/Science Photo Library, (right) P. H. and S. L. Ward/Natural Science Photos.

105, **Aziz Khan**. 106, Mary Evans Picture Library. 107, (top) **Aziz Khan**, (bottom) Science Source/Science Photo Library. 108, (top) **Aziz Khan**, (bottom) National Cancer Institute/Science Photo Library. 109, *Sir Alexander Fleming F.R.S., Discoverer of Penicillin* by Ethel Gabain. By courtesy of the Trustees of The Imperial War Museum, London. 111, *Vase of Flowers with Diamonds on the Table* by Jan Brueghel. Galleria dell'Academia Carrara, Bergamo/Bridgeman Art Library. 112, Glaxo Group. 113, Dr. G. Settles/Science Photo Library. 114, Don Siegel/Science Photo Library. 115, Godfrey Argent Studio.

The Cell in Evolution

116, *The Great Day of his Wrath* (detail) by John Martin. The Tate Gallery, London. 118, (top) G. Kinns/Natural Science Photos, (bottom) Natural Science Photos. 119, Ann Ronan Picture Library. 120, A. Hayward/Natural Science Photos. 121, By courtesy of the Trustees of The National Portrait Gallery, London. 122, Mary Evans Picture Library. 123, R. Royer/ Science Photo Library. 124–125, Tony Stone Worldwide. 126, Derek Berwin/The Image Bank. 127, *Sulphur Miners* by Renato Guttuso. The Tate Gallery, London. 128, **Mick Gillah**. 129, Mary Evans Picture Library. 130, **Mick Gillah**. 131, Ralph Wetmore/Science Photo Library. 132, **Mick Saunders**, (bottom) John P. Kelly/The Image Bank. 133, M. W. F. Tweedie/NHPA. 134, **Mick Saunders**. 135, (left) Martyn Page, (right) Dr. A. Brody/Science Photo Library. 137, M. I. Walker/Science Photo Library. 138, Andrew Mounter/Planet Earth Pictures. 139, Eric Gravé/Science Photo Library. 140, Dr. Ann Smith/Science Photo Library. 141, Eric Gravé/Science Photo Library. 142, Warren Williams/Planet Earth Pictures. 143, (left) Tony Stone Worldwide, (right) Ivor Edmonds/Planet Earth Pictures. 144, Flip Schulke/Planet Earth Pictures. 145, Mike Coltman/Planet Earth Pictures. 146, Tony Stone Worldwide. 147, The Kobal Collection.

Index